The Dominant Gene

A Novel Series about Becoming an Agile Leader

Francie Van Wirkus

The Dominant Gene is a business novel series about the journey to becoming an agile leader. At the highest level, an agile leader is fearless; to serve others, to learn daily, and to sense and adapt, all in the name of quality and customer value.

Each agile leader journey is different, because we are all from different industries, backgrounds and corporate cultures. No matter the road you take, you can be assured that the ride will be undulating, time consuming, and very personal. And, you will learn that the destination is not the end, but the beginning. The Dominant Gene offers an up-close view into one leader's quest to become and agile leader: the struggles, victories, and learnings. You're invited to join main character Joel, as he leaves comfort and complacency to begin his trek of uncertainty to become an agile leader.

I hope that you will entertained, disturbed, and motivated to end your own comfy world of complacency, and begin to grow. When it gets hard, when you feel like quitting, remember you are not alone; you have me and an entire agile community here to help.

Leading Unlikely

Book Three

Francie Van Wirkus

The Dominant Gene: A novel Series about Becoming an Agile Leader
Book Three: Leading Unlikely

Copyright © 2018 by Francie Van Wirkus
ISBN 978-0-692-063835

Edited by Kristy Gang, Jeff Shilling, and Claudia Marquette
Beautiful illustrations and cover design by Ian Corrao
Formatting by Polgarus Studio

For Claudia, the best partner and friend

for leading the unlikely of…anything.

I love what you cannot see:

fence lines, negativity, and the word no.

Contents

Acknowledgements

Thank you to Michelle Burns, for sharing your epic

Ironman Boulder experience with me.

I have new respect for your tenacity and your love of Mike and Ikes.

Once again, thank you to my lean-agile tribe of friends for

your expertise and support.

Let's celebrate this accomplishment, not be complacent.

There is still much to learn.

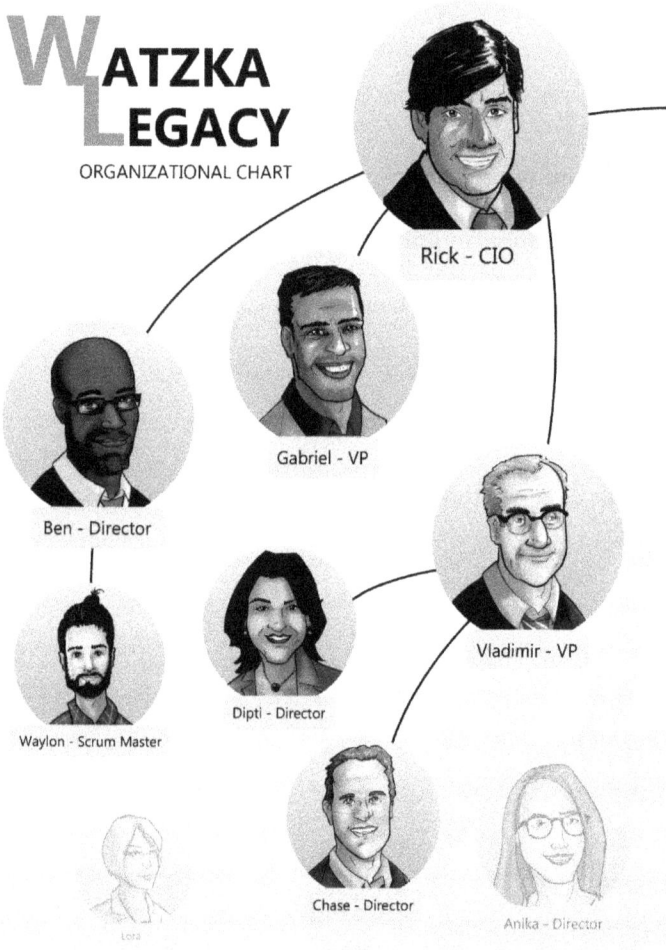

WATZKA LEGACY

ORGANIZATIONAL CHART

Rick - CIO

Gabriel - VP

Ben - Director

Vladimir - VP

Waylon - Scrum Master

Dipti - Director

Lora

Chase - Director

Anika - Director

THE DOMINANT GENE SERIES
LEADING UNLIKELY

Thad - VP

Eve - Agile Coach

Meryll - Director

Joel - Director

Jack - Director

Cindy - Manager

Vijay - Manager

Alexis - Manager

Karen - Team Lead

Chris - Team Member

My Vision

Professional
1. To be trusted by Lora and Rick.
2. Less wasted time on conference calls.
3. Fewer ultimatum days: You can do this, but then you won't have that.

Personal
1. Eat better.
2. Support my kids' events. I miss too many of them.
3. Fewer ultimatum days: You can do this, but then you won't have that.

Hello, friends.

Welcome to the story of my agile leader journey. I thought it would be good to set some context before we dive into things. My story involves a lot of interactions that should remain private, so I'm holding back the details of my work, my company, and my family. It's not important to know them to understand my journey, because what I have learned is in my heart and mind. I've changed the names of my family members to keep their privacy, too. On that note, you can call me Joel.

I live in suburban Denver, in the beautiful state of Colorado. I'm 46, and happily married to beautiful Cele for 19 years. We met after college, at a friend's Christmas party. We have four children, whom Cele is raising instead of teaching high school math. Caroline is 16 and looking for a job; she skis and plays basketball. But her greatest sport is hanging with friends. Elliot is 14, and into hockey and baseball. Eric is 8, and is living large. His latest interests are hockey and the violin. Then there is my adorable six-year-old daughter Cici, who wants to do it all.

I am an upper level technology leader in a large, complex, American corporation. Let's call the technology part of this huge company Watzka Legacy, or WL for short. I've worked there for 21 years. I'd like to say that's because I began my career when I was six, but that's not true. I began working at WL shortly after college, in my mid-20s, and grew through the ranks over the last 15 years. My title says *director*, but after the agile journey I have been on, I call myself a leader. More on that later.

Watzka Legacy has over 4,000 employees in the world headquarters alone. We sell…well, let's just say we create and sell an awesome product that requires lots of technology. In fact, technology is continually changing the game for our industry and for us. But this journey isn't about how Watzka Legacy needed to change. It's about how I needed to change.

Join me on my journey, which continues yet today. I don't have all the answers, but I have experienced explosive growth. It's time I share the value of what I've learned, so you, too, can become an agile leader.

Joel

Grow Happy, Grow Strong

"Well, that's one way to drop the bomb," Meryll mumbles to Jack and me as we sip coffee.

"You couldn't expect our annual vision and strategy refresh to happen any other way," I smile.

"It's not the news that was delivered," Jack is thoughtful. "It's that we're supposed to process the big news over a 20-minute break, and then go do something with it. That doesn't make sense. It's too much to ask."

Meryll nods slowly. "This wasn't planned, Jack. This annual session was on the books. Then they got the news about our competition's release of their product, Green, and our hefty regulation fines for the data security breech. We can't delay the strategy session because we're too inflexible for that."

"That's probably how our competition beat us out with Green in the first place," Jack sighed.

"So now it's our problem, and we have to whip up a real take-on-the-world strategy in response," Meryll rolls her eyes. "I'm sure there will be layoffs as a result of this problem."

I hold up a hand to my friend, "Don't go there, Meryll. We can't get all crazy about this without the facts." I can almost hear my coach, Eve, encouraging me to *lead the change one moment at a time.*

"Yeah, but there is just something about this news."

"Did Watzka Legacy lay off people in the past for problems?" Jack asks. "Of all the history you two have downloaded to me over the past year, I don't remember…layoffs."

"Not really," Meryll sighs. "But we have had people *affected* by problems, which is a nice way of saying *fired, or let go with a package.* All in secret, just five to 10 people at a time, so it's not noticed as much."

1

"Affected…" Jack's voice trails off. He's the new guy, chronically exposed to our culture by our stories.

I remember that *affected* period at WL well, and don't want to think about that today. "Let's save that for an office discussion. It's only going to slow us down with what we have to do today, anyway."

Jack, Meryll, and I commiserate often in each other's offices. Well, mostly my office. It's how we cope with dysfunction, and encourage each other.

Just then, the warning music played, signaling it was time to end our break, and move to our breakout session. No one in the break area moved very fast. The news that WL was upstaged by a competitor with a fantastic new product with a lousy name killed any upbeat mood we might have had attending an all-day strategy session.

At least for this breakout, I'm with Jack and Meryll. I'm confident on my own, and appreciate diversity of thought, but things are a little less depressing with them near me. Jack being the new guy, well, he's been at WL for about a year, gives us a fresh look at our work.

Chase, a director and peer of ours, reads to our breakout group from a sheet of paper, "Our goal for the next 30 minutes is to identify all the things that might get in the way of our new vision." He motions to the PowerPoint slide displaying our shiny, new vision: *grow happy, grow strong.* Chase has his poker face on, so as far as we know, he thinks this is a great time to talk.

I know him better than that. Chase is a practical guy, who cares very much for his teams. He's probably struggling to get through this session as much as we are. I hope the group goes easy on him.

"These enablers are the things that are the engine behind *grow happy, grow strong.* They are: customer focused, workforce engagement, agility, and global tech leader. Our enablers are driven by offering the very best customer care, having the very best product available to our customers, and creating the very best ways to sell it to them. The outcomes of these enablers and drivers are loyal customers, an engaged, digitally advanced workforce, and continuing to be an industry leader…"

By now everyone is beginning to lose focus. Not because the vision and its accompanying enablers, drivers and outcomes are overwhelming, but because our minds are still spinning with the news that a competitor took a giant bite out of our market share. And because we were heavily fined for a regulation snafu. A double punch that makes all of us very nervous. The questions swirling in our heads are drowning out everything else.

"Excuse me," Rick enters the room. "Joel, Vijay, may I have you for a few moments?" Rick continues to apologize to Chase as Vijay and I leave our tables, and walk toward the door.

"Come with me," Rick leads us to an empty breakout room.

WL incurred a $10 million fine for storing personally identifiable information (PII) about customers online without adequate security safeguards. An estimated 300,000 customers were likely exposed. Names, addresses, credit card numbers, and bank account numbers. Holy crap. This is more than a regulation snafu. I figured it was something about protecting data, but I didn't think it would be so close to the work Vijay's teams do. Well, it *is* what they do.

A cloud storage service had a misconfigured bucket, which apparently was a third party contractor's fault for not properly securing the web service. We're not the first company to have this misconfigured bucket problem, I guess. To make things worse, and where the fine comes in, is that apparently, we knew it wasn't right, and didn't ask the third party to fix it.

To make things even worse, the news was leaked to a national technology news source before leaders found out about it here. A well-known cyber security reporter then blasted the news on their site. It rapidly spread from there. Our company is now in damage control mode on many fronts.

It would be easy to get lost in the details of the fine or the investigative reporter who learned of the breach, but Rick wants us to instead focus on how WL should store PII in the future. He believes the FCC's action was a signal to WL and companies like us that they are looking at our industry more closely. They are taking the Federal Trade Commission's lead on making companies take industry appropriate steps to protect some kinds of PII. WL attorneys are looking into whether the FCC can use its authority like it is, but we can't wait for that to be resolved. We are being watched, and need to fix this.

"We have a data security item in our backlog," Vijay says. It's been a known problem for a long time. For years, in fact, Vijay's team has been asking to buy a new big data cybersecurity platform. We've needed a single, comprehensive view of business risk. A way to see relevant threat data, and then develop ways to address them. But the idea always got knocked down on the priority list. Something else always came up that was more on fire, and more important than a nonfunctional security requirement.

"We just made some headway with a cross-functional team on this. They have some fresh voice-of-the-customer feedback, and potential solutions lined up. They have been building relationships with a few, select vendors, just in case their work is ever prioritized higher than usual. It would be good to get them involved now."

Rick waves a hand, "None of that matters now, Vijay. We can't wait for anything or any team. Not even our customers. This needs to be tightened up. Fast."

"I understand," Vijay looks down. A deflating moment for us, and more importantly, our customer's data.

Rick is in job preservation mode, clamping down on us. Right before our eyes, we are losing any ground we had transforming WL from a command-and-control, hierarchical company.

No doubt WL has a terrible problem on its hands, but Rick is reverting right back to old ways of leading. Dictating. If I say something about it to Rick, I could cause problems with him. If I say nothing, I will be a sellout to Vijay. And to all the coaching I've had from Eve.

"I realize I have been one of the people telling you and your team no, Vijay," Rick sighs. "But we are in deep, now, and have to take immediate action. I have a tremendous amount of pressure on me to quickly remedy this, and prevent it from ever happening again."

Well, at least he's admitting he's reverting to his and WL's old ways.

Vijay looks at me through hooded eyelids as if to say, *aren't you going to do something?* He's right. I need to try to change Rick's mind so that he sees the value in working with the teams closest to the work.

"Rick, we understand the need to shut this breach down, and immediately prevent it from ever happening again. You must be hearing about it from all sides."

"Tell me about it," he rubs his temples.

"The teams closest to this work need to be involved, Rick. They have deep knowledge of the systems, the voice of our customers, and knowledge of how a new cyber security platform can interface with it. No one at WL knows any of this better. This work wasn't finished because it wasn't made a priority by WL, not because the team is slow or bad, or can't be trusted."

"You have been bending my ear about this for a very long time," Rick touches Vijay's sleeve for a nervous nanosecond.

"We can't shut the team out; they know the work better than anyone else. We need them, Rick. Our customers need them."

Rick sighs. "I know what you're pushing for here, Joel. But I am going into two status meetings per day on this problem. One from the regulatory angle, and one from the fix angle. I can't just leave this to the teams. I have been given four director's time to help me with this. A small team, if you will.

"We have a stand-up each morning," Vijay offers. "This is when the teams talk about what they got done, what they plan to do, and what is getting in the way."

"Are you saying I should attend that meeting?"

"Not exactly," Vijay is uncomfortable saying the plain, old word, no.

I ask Rick who the directors are, and none of them are from tech. They are from the business side. Of course. *They* were going to be charged with buying us a big data cyber security platform? Classic WL.

I tell Rick that the team's standup is for the team, and that if anything gets in their way that they can't solve on their own, they will ask the leaders or stakeholders to help solve. How about we pull the teams together; Rick and anyone else with vision on this work can come and talk with the team. Make this the absolute number one priority, and give them whatever they need to get it done. Do not ask them to deliver anything else but this single priority.

"There needs to be a tight feedback loop between the team and the decision makers, so the product is made to the exact specifications. This is the heart of scrum. As often as needed, the team can show any stakeholders what they made as they go along. Normally it's every two weeks, but in this case, we may need to show the team's work daily, for a while."

"You can do that? Show us what got done on a *daily* basis?"

"Only if it makes sense, Rick," I say. "Every time the team engages your stakeholders, they are making your product better, but they are also taking time away from work to do this."

"How long do you think it will take? How long for us to get this platform in place?" Rick asks.

Vijay said it's hard for him to say, and that the best thing to do is to ask the team. They can't make a promise, but they can estimate the work.

"Everyone can't work 80 hour weeks. We probably need to throw more bodies at the problem. Joel, you can hire more resources, er…people if you need to."

"Rick, you've had a lot of experience leading projects," I say.

"Yes."

"In your experience, what happens to the work when we add more people to a project?"

He sighs, "It slows down. I got you, Joel. Thank you. There is just so much pressure on this from corporate."

I assure Rick that if we need more people, and the benefits are stronger than the risks, we will add people.

"Okay, I feel better about this. At least, about fixing it. I'm willing to give our expert teams the chance to help dig WL out of this hole. Who knows, it might be a great example for us to use with future transformation. But I'm getting ahead of myself. You two should go back to your breakout session. But tomorrow, we need to meet up first thing with your teams."

As we walked back to our breakout session, Vijay is on his phone with Rick's assistant, to get him booked for a two-hour discussion with his team. The other directors assigned to Rick will also be in attendance. Then, he calls one of his team leads, and gives him the basics of the team's new priority.

I'm very tempted to make other phone calls to related team directors, but that will only create more swirl for Vijay's teams. Instead, I offer my support to Vijay. He thanks me for practicing what I preach. We high five, and then walk back into our brainstorming session.

We're welcomed back, and asked to just fold into what's happening.

"We'll brainstorm the things that get in the way of realizing our vision, and talk about them together as a group. Then we'll report out to the larger group."

"What does that mean, really?" Dipti, another director, asks.

I hear Meryll sigh, and a few others shift in their seat. Here we go...

"Good question, Dipti." Chase is the professional, even though he probably wants to throw up his arms and say, *I just told you what it means!*

"And why aren't any words in this vision capitalized?" Another director huffs. "All lower case looks so dumb."

"We need to focus first on our enablers. These make up the engine behind *grow happy, grow strong.* None of the other parts, the drivers or outcomes are going to happen if we don't get enablers right." Chase takes a breath. "They are: customer focused, workforce engagement, agility, and global tech leader."

Dipti takes notes and nods her head. The room remains silent. It's so flat, Chase has to notice.

And he does notice. He adjusts his shirt in obvious discomfort, and sits down, "I don't know how all of you are feeling, but I want you to know that the news we received this morning is making this very difficult for me. If any of you are feeling the same way, I invite you to try to relax and let go of it for now, and let's just get through this exercise." Chase offers a smile. "If nothing else, enjoy the irony of it. Here we are, struggling with news about new problems for WL, and this group is organized to look for the problems with our new vision."

The room is still silent. Then Meryll offers him her appreciation for being real, and suggests we all convene at the bar later to talk through it. We laugh and the room mood is elevated, even if only for a short moment.

Chase is bothered by all of this, too. Of course he is. Eve would be proud of him for acknowledging where people are at in the room, instead of acting as if there is nothing wrong, and just pushing through. But Eve is only my coach. My agile leader coach.

Deliver. Drive for results was the old saying. Well, it still exists, just not in my mindset. Those words aren't bad; the senior leader behaviors behind them are. This pressure is the usual way people are crushed at WL.

Several months ago, Jack, Meryll, and I were forced into coaching by my former supervisor and vice president, Lora. At first all of us hated the idea. I thought the coaching was going to force me into a methodology that I didn't understand, let alone believe in. That didn't happen, and I've learned that my coaching experience with Eve has been the best thing to ever happen to my career. Jack and Meryll feel the same way. In fact, Meryll's entire life was impacted by her coaching experience. She learned things about herself, and has since ended her marriage with a spouse who gambled away most everything they had, lost so many pounds, I can't keep track, and is all-around living a far healthier lifestyle. I've gone from watching my peer and friend morph from burning up in front of me to becoming happy and healthy again. The power of this coaching is not in learning tools or what to say, but personal growth. From that, the potential to change feels unlimited. Incredibly hard to do, but unlimited upside.

Lora left mostly a legacy of being a control freak with a poor leader filter, but her idea to send us to coaching was brilliant. In fact, our CIO Rick continues to sponsor our coaching engagements, though he hasn't done it himself. He never talks about Lora, except to give her credit for introducing coaches to our organization. Maybe someday it will be expanded to other directors, but for now, I think Rick wants to contain the experience to three people. It's easier to keep track of just three transformed people.

My peers and I brainstorm ideas using flip charts. Jack is our scribe. He beat me to the punch volunteering for the job. He's catching on well. Those of us with years of experience at these events know that jobs like discussion facilitator and scribe make the time pass quickly, and you are usually held to a neutral position. This is great, because no one wants to commit in these discussions. And yet we are encouraged to think outside the box. *Get creative! The sky's the limit!* Then, later you will be pulled aside and asked to explain why you were attempting to *steer the group in a radical manner away from the direction WL is taking.* You might even be put on an action plan to study the WL strategy and drivers more closely, so that you can *stay current with the company's direction.*

Yes, in this strategy session, we know better than to be a voice for change. So, we will offer the most executive-pleasing, vanilla responses we can. Be engaged, but don't rock the boat. Because we don't really want to change.

Dipti, another peer, offers that culture will hold us back. It hits a hot spot with the group, and a lively discussion ensues. Well, as lively as it can be after the big news about WL. I'm not sure I agree with the idea, and I'm not sure why.

"If we can change the culture, then those enablers of customer focused, workforce engagement, agility, and global tech leader will happen," Dipti is energized. "It's what's been holding us back the entire time. And isn't that why we bought Agile and Lean to WL, to change the culture?" Dipti holds up a hand. "I know, I know, we shouldn't say the word *lean*. My bad. *Project Learn*, the umbrella of all our digital transformation efforts, is the right phrase."

Dipti made the same misstep most of us already have, ever since the naming of our transformation by executive leaders who decided WL was going agile, and beyond. The *beyond* part meant that the company was committed to a path to grow, learn and transform into this new way of working. They wanted to *name* the big picture of process improvement at WL.

Project Learn is just too close to *Project Lean*. Eve convinced me that I didn't need permission from the executives to change the name. If our team is leading the transformation, then we can drop *Project Learn*.

"Good catch." Vijay, one of my managers weakly offers. Eternally positive and upbeat, even Vijay is a little off today.

Although no one else is, Dipti is on a roll. "Our culture holds us back from so many things. Even the way we run these vision and strategy sessions."

I don't recall Eve ever coaching me that agile or lean were designed to change culture. They are both methodologies, mindsets, but culture changers? Do companies use agile and lean to change cultures?

In the past, Eve and I have discussed WL culture, but there is something about this conversation that has me curious. But not enough to speak up. I don't want to share my thoughts with the group, because aside from Dipti, the mood in this room is in no shape for any sort of philosophical discussion. *Drive for results* looms over us. I will definitely have to kick it around with Eve at our next meeting.

Meryll agrees, "Yes. If only the culture would change, these enablers would work. Well, I don't think they are the greatest enablers, but I guess that's what we have to work with. And, I remember Rick talking about culture at our last leadership rally."

"The one where Rick told us about a mistake he made in 1993." Jack puts an index finger to the air.

Meryll gently claps her hands together, "That's the one!"

"I think it was supposed to motivate us." Chase tries to remain in neutral facilitator mode. "He said the way we lead needed to change, and that we should look at mistakes as learning."

"Yep, that's what I was thinking." Meryll sighs. "That was months ago. We haven't heard anything on the topic since."

It's as if these two are the only ones in the room. Not in an uncomfortable way. Likely, part fascination that they can recall the messages we've heard, and then there is our own distraction about the security breach. How will this change us? What about our teams, our work?

"Our VP Thad has mentioned culture, so that's not entirely true, Meryll." Jack says. "But his message was different. At least, that was my interpretation. And listen to us," Jack chuckles. "We spend so much time interpreting our leaders. Wouldn't our culture be better if we weren't wasting our time interpreting our leaders?"

"A stiff head-wind of culture." Meryll proudly announces.

"What?" Jack squints. "What does that have to do with wasting time interpreting our leaders' mysterious ways?"

"Thad." Meryll looks around, double checking that it's safe to discuss our newest VP who happens to be Lora's replacement. "In a team meeting. He gave us this big u-rah-rah speech about how we have to make changes and *culture* is the stiff head wind we face."

Jack's eyes dance, "Ah, right! The metrics schpeel."

Meryll lowers her voice, offering her best Thad imitation, *"We have to drive forward with solid metricsssssuh."*

"Bullseye." Jack and Meryll high five. I shake my head with my *you two crazy kids* smile.

"Well," Chase wipes the silly grin off his face. "How about we get back on track?"

As if there was a track. This room is so checked out, it's a wonder we are all here and not out in a hallway on the phone, talking with our teams. Better yet, meeting with them to digest the news. Digest it, process it. But no, we are asked to drive forward. Drive for results. Lead the change…

Dipti clears her throat, "So with agile and lean, our culture will change for the better. Is everyone in agreement that culture can be our focus for today's session?"

It wouldn't be a WL meeting if someone didn't ask to clarify things further. Even in the state this group is in, someone will do it, because we've all been conditioned to do it. It's part of our painfully slow consensus culture, revealed to me by Eve. In this case, Vijay leads the charge with the classic questions. "Do you have a sense for what that might look like? And, how can we leverage this at WL?"

Chase stammers a bit, and checks his watch. We have 15 minutes remaining. "Uh, those are good questions, but we should really stick to the question we have been asked to answer."

Vijay reads the PowerPoint display again, "Identify all the things that might get in the way of our new vision: *grow happy, grow strong.*"

"Right," Chase nods. "And we have 15 minutes to do produce our answers."

Meryll is on it. "Look, we're all distracted, but we can do this. We don't want to be the group that couldn't come up with anything. Then we'll have to work on this outside of this event. Anyone have time for that?"

Quickly, the group rallied, and created a short list of what might get in the way of our new vision:

Culture
Outside disruption
Lack of funding the right work
Legacy systems with technical debt

A solid list of vanilla corporate impediments. No one can argue them to be untrue; they sound challenging, and are known company-wide. The last thing this group wants to do is introduce an impediment that no one has ever heard of. That will create uninvited chaos which only leads to more work than what we've already invested. It would start with questions such as *When did this become an issue?* and *What does that look like?* and even better, *Well, I hope there are some great examples of this problem.* My personal favorite is when someone asks *Didn't we try to work on that once before, and it failed?*

Our list is submitted along with those from the other breakout sessions. We are a sharp group and know how to produce stuff when it really matters, even when we are run down and deflated from bad news. So many times, we have created a list like this. And mostly nothing is done about it. Well, we try to do something about it, but there are no *outcomes* from it.

Normally, a committee is formed with the goal to further explore what should be done. The committee means well; like most people at WL, but then there is the reality of WL. They will meet for an hour a week, or two hours a month. Sometimes an idea or improvement makes it through, but most of the time WL crushes it. This can happen quickly, when funding is decided late in the year. The big improvement we had to have is now determined not to be important because of some other shiny object. Although a little dejected, most on the committee breathe a sigh of relief that they no longer have this work. Not because they didn't believe in it, but because some pressure has been relieved. Most committee members have so much WIP, work in process, they have to work off hours on committee work.

Or, the work will be crushed slowly, a *death by committee* of sorts. The people on the committee don't (usually) kill the work. Executive leaders lose interest in it, in favor of the next distraction, so funding is siphoned away, or cut off all together. Sometimes committee members with access to a budget will find a way to keep the work alive, pillaging their own budgets for money and cobbling enough together to keep the work alive. Picture three or four directors walking into a candy store, with a large jar full of change. They give the store clerk their order, and then dump their change onto the counter. The executive leaders don't mind when this happens, as long as their funded work gets done.

With multiple players on the committee, it's hard to make everyone happy. Often, the ones who make the biggest fuss about the date and time of the meeting are the first to stop attending. There will be shifting of the time and date of the meeting because the people on the committee are busy and can't make this effort a priority. Calendar entries of cancellations and reschedules will overflow in my email box. Fortunately, my assistant Marilyn handles most of that chaos.

Then, during the busy months of November and December, the committee will still meet, and be thinly attended, or the meeting will be canceled all together. In January, the meetings will be reset, usually with a new facilitator who has been asked to breathe life back into the committee. The facilitator accepts the challenge with the best intentions.

With likely three months passed since the last committee meeting, it's difficult to get started again. The purpose often gets foggy. People who originally started the committee notice that the intent is not the same, but they're usually too busy to do anything about it. And, they don't think they can stop attending the meetings because it will make them look bad.

We all know what will happen, or not happen, after our list is submitted.

On our next break, the three of us are quiet. There was so much dumped on us and no let up with the breakout session. At this point, I can only promise myself not to do the same to my teams.

"Well that was refreshing." Meryll breathes.

All I can do is give her my *just-another-day* look.

"Joel," Meryll touches my sleeve like she does when she has a deep thought, "Dipti's comment about Project Learn reminded me that you haven't been announced as the new leader of the transformation."

"Yeah," Jack chimes in. "You're the Agile Transformation Sultan. When are you gonna start throwing your weight around?"

"Yeah, I have a final review with Thad coming up. It's business as usual until that happens," I say.

Meryll scrunches her nose, "Weren't you supposed to have that meeting over a week ago?"

"Actually, two weeks ago," I say.

"Well, that's ridiculous," Jack says. "No one should keep the ATS waiting. Ever."

"ATS?" Meryll and Jack are back at it. The best thing for me to do is let them kick this around. I'm annoyed with Thad being a bottleneck to my work, and this banter helps take the edge off.

"Agile Transformation Sultan." Jack affirms.

"Ooo," Meryll quietly squeals. "That sounds powerful. And yet, here we are, visiting with the ATS himself. He seems pretty regular."

"That's the problem, Meryll. He puts his pants on like everyone else. He should really start throwing his power around more."

"Right on, Jack."

Grinning, they high five each other, and then turn to face me.

"You're killing me," I smile.

"Why did Rick pull you out?" Jack asks.

I share the basics of Rick wanting to go in command-and-control mode over the remedy for our regulatory sins. I share more about the data breach. Neither of them are surprised.

"How ironic," Meryll chuckles. "That can has been kicked down the road so many times. Now, it's kicking us."

I sigh, "I can't switch from crisis mode leadership to *grow happy, grow strong*."

"We only have a few sessions left," Jack encourages. "We got this."

"What do we have?" Meryll asks. "More and more piled on top of us, and consequently on our teams."

"As long as I don't have to go on a culture committee, I'm all right." I say. "And, I will protect my teams as best I can."

"Right." Jack and I clink cups.

"This is too much." Meryll shakes her head. "I am trying to let go of it and just focus, but I can't."

"None of us can, Meryll." I say. "As I see it, there are two options: leave now, and go reflect on what you heard for the rest of the day, so you can get ready to face your teams; or stick it out as best you can, knowing that the quality of your experience here is way below your standards for this kind of event."

Jack raises his eyebrows, "Wow, you've been processing this."

"Fight or flight, Jack."

"It really is a fight to remain here," Meryll sighs. "I knew those were *my* options. I hoped maybe *you* had some other option, but I see now that we all have the same problem. And, I hate these options."

Jack raises his water glass, "To options."

We clink our glasses together, and then it's time to head to the next session.

For the close of the meeting, there was an executive interview. The chief financial officer was interviewed by a sales director. They started out chit-chatting. Literally. It was obvious the director was interviewing the CFO. This might have been an acceptable get-to-know-you sort of thing, had we not just heard big news that impacts our company. We have no idea what's next, other than we have a new vision and a new focus.

So now we all found ourselves waiting and waiting for the small chat to end so the CFO could get down to some important message. Is there a shoe about to drop, or is this going to be a *keep calm and carry on* message?

Once they finish talking about the CFO walking his beagle, every evening to clear his head, we hear actual business data. The company is in a strong position to weather this storm, but we must take a hard look at our expenses. We will be following a strategic expense management approach in the near future. We don't have all of the details worked out yet. We don't have to do anything different today, but soon areas of the company will be contacted to discuss how they can contribute to expense management.

One area that is ahead of everyone is the tech sector of our company. They are *going agile*, and as a result will be working in new ways. They will be delivering technology to our customers faster, and for less money. This is the moment when Jack and Meryll look at me. I'm focused on taking notes, so I don't make a ridiculous face, and then get called out for acting like a sixth grader. I'm supposed to be a calm leader, even though I want to stand up and shout: *Agile doesn't cost less! Agile won't make us faster! We are transforming an entire organization, and that will actually slow us down for a long time!*

The CFO closes with confidence that we will *"come out of this speed bump stronger,"* and we *"will have learned new things about ourselves."* Everyone claps and then we are asked to watch a video. A *grow happy, grow strong* video plays for us, with uplifting music and all manner of fanfare. After the news we just heard, maybe this is a good idea.

We are bombarded with messages about how WL is a great place to work, and people are really happy here. Maybe it's not a good idea. Actually I don't know what to think. This is because I'm officially overloaded for the day. I'm a positive guy, but I can't soak up another message or slogan. So when phrases like *Try new things!* and, *Think outside the box!* appear, they fall flat with me. I don't know for sure, but it looks like I'm not the only one. Thankfully, it's a short video.

The sales director is back on stage. To help kick off our new vision, *grow happy, grow strong*, we are launching a new line of spirit wear. Just then, the CFO, who is stage left, tosses the sales director a trucker hat emblazoned with *grow happy, grow strong* across the front.

The director points to his head, "How 'bout it, eh?"

The audience claps.

The CFO calls out that the crowd is jealous, and the sales director agrees, all the while hamming it up with his new hat on. The CFO walks back on stage with a giant box of WL trucker hats. They are overflowing on the top.

The sales director makes a joke about strategic expense management, and then it's announced that everyone in attendance today will receive a WL trucker hat. Employees will receive a t-shirt, and can wear them to work any time they like. We are, however, advised to use discretion with meetings, as always.

"That was rich." Jack says as we walk out of the building for the day.

"Jack," Meryll growls, "remember the 50-yard rule."

"Who cares? This thing is an absolute circus anyway." Jack turns his trucker hat backward.

Meryll huffs, and reaches over to turn Jack's hat back around. "Maybe so, but I don't need any more negative feedback."

"You two should just be happy you have WL spirit wear," I chuckle. "I feel like I'm in high school."

Jack raises his arms up, "Next thing you know, we will have WL pep rallies. We should have them at a bar."

"Don't even…" Meryll sighs.

"Well this was quite a day, kids. I'm sure there's something good about today. I just don't know what it is after *that* extravaganza."

Jack playfully punches my shoulder. "At least they mentioned you're going to save the company a lot of money, and deliver stuff faster. No pressure."

"I don't recall my name being mentioned, Jack." I'm too darn tired to play along.

"Agile," Meryll sighs. "They are grabbing onto a methodology and making it the answer. The answer for things to change, the answer for things to turn around and get better, and the answer of what to blame if things go poorly."

My reflections notebook is going to be full after this event. Reflection is part of the lean leader standard work that Eve has shown me. Since our first coaching session, I've been asked to use a notebook for my learning reflections. Using this notebook is serving two purposes: first, to record learning thoughts to discuss with Eve; second, to practice actually doing reflection.

At first I was worried about keeping a journal, because I didn't think there would be value in it, and because I thought I would have to write a lot to satisfy my coach. Fortunately, Eve wants to cut to the chase with what I'm learning and thinking. Her guidelines are to keep my reflections to one sentence, maybe two. I could buy into that.

As much as I liked not having to write a lot, the short sentence approach is not easy either. It's very challenging to compile all of the meaning behind what I've learned about something in one sentence. One sentence feels so liberating and yet meaningful. No spinning it up for the sake of volume.

Over time, the practice of reflection has really grown me. I see the value of capturing my learning, especially on a journey like this transformation. Things move so quickly and change so fast; my journal gives me at least one way to capture my growth.

The other benefit might seem obvious, but is challenging for me: the actual *act* of reflection. So often, I go from one thing to the next. I know I should make time to stop and rethink, redo, or reconsider; I'm smarter and more creative when I do this. Calmer, too. But I don't. But over time, I've practiced enough that I have a fairly strong reflection habit. So much that when I'm in an event like this one, I feel compelled to start writing notes during the day, so I don't lose the thoughts later.

With the changes happening at WL, and the news we heard today, reflection might be the only way to keep up.

Reflections

Reflection makes me a better leader.

I need time to think about change and bad news. I can't just accept it and move on to the next thing.

Culture - is it a practical scapegoat, or an impediment we can actually fix, change or remove?

Agile is being positioned within WL to save money and help us go fast. It won't do either.

Report Out

It's the second report out in a week for our very small transformation team. It feels like great progress. Just a few weeks ago I hit the SEND button on my laptop, agreeing to Rick's request to lead the transformation of WL's tech work.

Too disheartened over the years and recent months with our company and its leaders, I didn't have the appetite to lead something like this. It seemed impossible to lead. Plus, the change was already out of the gate with some stigma, because it was poorly introduced. With executive fanfare, people were told *We're going agile*, and corporate communications named this work Project Learn. The name was easily mistaken for Project Lean, which, by the way, is a forbidden word at WL. I had pushed corporate communications and Rick to be transparent and call it lean if we are doing lean, but neither would accept the change. Also, the senior leaders had this myopic view of us going agile only impacting tech areas. Our business side didn't know about it, and if they did, they didn't have any exposure to it to gain understanding. My rant doesn't even touch the dysfunction and missed alignment within my own leadership team when Thad was hired.

My coach, Eve, wasn't impressed when I brought these concerns and facts to her. She and I both knew I was capable of leading this massive change; but she believed I had the wrong perspective for it. I was looking at all the problems as if they really held me back. I didn't think I could really change any of it. After a few years of getting beat down, I lost my mojo for such thinking. It was impossible for me to lead a change that I believed was impossible to actually happen.

As only a great leader coach can do, she helped me see that I wasn't giving myself permission to lead past any of it. She reminded me I had everything I needed to lead this transformation. I had the ability to change the name of Project Learn to something better, or maybe not name it at all. I had the ability to face Rick and tell him we will use the word lean. And stand up to any other leaders or stakeholders who preferred to hide from the truth. I can make sure our transformation strategy involves our business partners from the beginning. Practicing agile at the technology level would be okay, but we have to involve everyone if we want to transform the organization.

She also convinced me by a very thin margin that all the leadership dysfunction and lack of alignment on my team was only in my way if I allowed it to be. Eve wasn't telling me to bust through it all, and not care about the consequences; she made me see that I don't need permission to lead through it. She made me see that all of it was entirely possible. That's how and why I agreed to lead the transformation. When I gave him my acceptance of the job, Rick was thrilled, and said I had his full support.

So I pulled together a small team to get started, consisting of one of my managers who is very open to change and experimenting, Vijay; Ben, a director who has the first three scrum teams under him; Karen, a consultant who I've worked with in the past who has scrum master experience from outside WL; and a few of the scrum team members from the three pioneer teams, including Waylon, the scrum master from team Can't Make This Up.

The goal of this team is get us to the point where we have a vision, a strategy, and a basic roadmap of how we'll get there. Having a narrow focus helped our team work together very well; we all have great energy for this change, and with just a few of us, we could make decisions quickly.

We didn't make a decision in a vacuum, even though it feels like our senior leaders just decided one day to *go agile*. They had some urgency: we need to make things that are technologically relevant to our customers and we need to do it with business agility and customer focus. What we learned at our recent all-day meeting reinforced that need. Still, the transformation team needed to know what the tech teams and business partners and teams thought. After all, they are the ones who will carry the majority of this change. This is how we can actually transform the entire organization.

We got right to work on surveying the tech and business areas, to understand their needs and the opportunities around us. We called it our *feet on the street* research. This work was extremely labor intensive and time consuming. But it gave us incredible insight into what was going on. Ben called it harvesting the opportunities and needs of the organization. It made more sense than *feet on the street*, because it sounded like we were going to actually do something with the data, so we have adopted Ben's words.

We also conducted two formal surveys. This was a strategy Vijay encouraged us to use, so that we could have a thorough set of data, and to appease leaders. We all knew the *opportunities and needs* data we harvested was real and current, but our senior leaders are very numeric driven. Vijay said he learned that what they have is quantification bias: a mindset that doesn't allow for non-numeric data to be used to assess organizational health. In other words, a belief that the only data with integrity is numeric data.

I think I had that bias too, until Vijay taught the team about it. People, mostly leaders, have a strong desire to use numeric data for everything, even for measuring organizational health. Striving to become a data driven organization (Vijay said all the cool-kid companies are doing it) only strengthens the misconception that having actual conversations with people about change is not real data. Why are people like this? Because they choose to be, or because they lack the intuition to consider and use conversational data.

So we had the company's vision of *grow happy, grow strong*, we had conversations, and we had numeric data. This provided us with a nicely rounded view of WL's organizational health. We had what we needed to feed the vision for transforming WL. It was exhausting work, but our small team was motivated to make this vision happen. And so it did, and now we have an audience to hear about it. We shared the high-level trends of what we learned:

- Everything is a #1 priority
- We have too many projects going at once

- We are order takers from people far away from the work
- We waste time in meetings and reports
- People are hungry for change
- We have convoluted and broken processes
- Leaders are roadblocks to progress

"Welcome to our transformation team's report out. This is our second planning event in two weeks, as we work in earnest to shape the transformation of WL's technology work. Today, we will share with you our vision, our foundation upon which we will grow."

Vijay continues to set the stage for the report out by reminding us of our corporate vision to *grow happy, grow strong*. Once again, those enablers are shared: customer focus, workforce engagement, agility, and global tech leader.

"As we finalized our focus, we were very inspired by the way the agile manifesto was written:

Individuals and interactions over processes and tools.
Working software over comprehensive documentation.
Customer collaboration over contract negotiation.
Responding to change over following a plan.

If you're not familiar with the agile manifesto, that's okay. You can research it later. For now, when we all see this on the screen, we see one word common in every sentence. That word is *over*. My teammate Karen will now build on this."

The room feels full, but it's a small group of leaders and most of the three scrum teams. We invited several business partners, and one of them showed up. Rick is there, along with Ben, a few other directors, Meryll and Jack. Thad didn't attend.

The team believes that the word *over* in the manifesto means that there are no absolutes. There is guidance and a foundation to stand on, but it is not an inflexible rule as to how people should change. There is organization of intent and thought, but not a directive on the precise way to be an agile team.

This philosophy resonated with the transformation team; they want to be a guide, a foundation for change at WL, but not script inflexible rules. We don't want to centralize ourselves to control the organization. We see ourselves building guardrails for our teams and leaders, so they don't fall away from us, but giving them a wide road to work and discover how they can change. We are here to lead the transformation and help, but not dictate it.

Also, when the agile manifesto was written as *We are uncovering better ways of developing software*, the founders positioned their methodology without an absolute. They knew there would probably be better ways to make software in the future than the method they designed: agile. In fact, they wanted people to use it, and make it better. So, we want to do the same with how we work at WL. We don't see our transformation as an agile transformation or a lean transformation. It is a *learning* transformation.

"Given that context, and given the overall corporate vision, we want to show you how our vision evolved with the team. It was a three-part iteration." Karen shows us a wall with giant sticky notes on it:

Sticky note 1: We are using new and better ways to create and deliver software, so that WL can remain the customer's choice.

Sticky note 2: We are continuously learning how to better create and deliver ways for our clients to connect with our products.

Sticky note 3: We are continuously learning how to better create and deliver ways for our customers to connect to the world.

This is when Thad appears, sits down in the back of the room, and opens his laptop.

"I like what you did, here," Rick smiles. "I agree in staying away from putting a methodology in your vision. I also like your nod to the agile manifesto. This makes your vision strong, and sustainable."

"Do you have a sense for how the rest of the company might react to this vision?" Thad asks. He was in the room no more than 90 seconds before speaking.

Karen takes on the undefined question with a smile, "Please share with us what you're curious about, Thad."

"Did you survey your customers, to get their thoughts on this change? I hear you say customer focus, but I'm not seeing where you got the customer's voice."

Vijay tells Thad that two different strategies were used to gather data: harvesting opportunities and needs through conversations and formal surveys. Although it was already announced that *we are going agile*, to help the company remain competitive and create relevant products, the team spent hours engaging WL employees in tech, and in the business. They also surveyed business partners. They did not talk with every single person, but with many. Vijay points to a giant sticky on a side wall that had a printed heat map taped to it. This map represented the parts of WL that were surveyed, not names, but areas of work like engineering, information risk and cloud.

The team also hosted lunches, listening sessions, and even an open mic session, where business partners could share their wish lists for improving tech. It was an incredible experience for the team, and helped the team see beyond agile, DevOps, or any other methodology, to the big picture of how WL needs to change.

"But that's just anecdotal evidence," Thad says. "I would think a survey with actual data and numbers would be the best way to survey the group. After all, we are working on becoming a data driven organization."

"Yes, we are. The content we harvested is good data, Thad. Our team has learned about something called *quantification bias*, which studies have shown holds an organization back from its potential. You see, leaders are so very used to having numbers to drive our decisions, that anything that is not measured this way is either dismissed or overlooked. Leaders want and love their numeric data."

We didn't think we'd need to share that, but Vijay has read a ton on the subject, and was certain someone would challenge the way they collected data. This is the power of our team planning this report out together. I will have to remember to praise Vijay and the team for all of these little moments that add up to a well-prepared discussion with our stakeholders and future customers.

"I wasn't dismissing this information; I was saying it can't impact your decisions."

No one knows what to say to Thad's comment, so they say nothing. Vijay ignores it, and jumps back in.

"So, we are striving to look intuitively at the opportunities and needs in our organization. We intentionally looked for feedback from teams because they will most likely be the hardest hit by this change, whether they are on a scrum team or supporting one."

Vijay talks about the two formal surveys that were done over the last several weeks. Thad asked about the response rate, which Vijay said was about 56 percent. A healthy response rate, but the in-person conversations had a 100 percent response rate. Vijay has a wise way of working when in the hot seat. He described how the team used all this data to help drive the transformation, and they reported out on it after the last planning session. Vijay knew Thad didn't attend the last report out, so he gently directs him to a wall in the back of the room, just over Thad's shoulder, where all the data is summarized.

"What about middle managers?" A business partner asks. "Were they surveyed? There are so many of them here, and from my experience, they have their hands on all the work."

Vijay and Karen look to me, although they know the answer. They want to see me lead the transformation, so they can do it, too. I don't need a formal announcement to do that, either.

"Great question. Normally, we do implement many changes though our middle managers and people just like me. But we've been talking with organizational change experts here at WL, and their research shows that middle managers are often one of the greatest impediments to agile adoption and overall transformation. Middle managers are not inherently bad, or wanting to sabotage change, but our environment sets them up to actually be roadblocks: things like funding, layers of project hierarchy, and corporate hierarchy. We're not leaving them out of the change; we're just not beginning with them. We could geek out on that right now, but let's keep our focus on this report out. I'm happy to discuss this with you later."

The business partner was satisfied. I was sweating. Hopefully no one else knew it.

Rick changes the subject. I guess as the CIO, he is allowed to do that. He likes the customer focus topic, and urges all of us to keep it top of mind. This means we will have to constantly work closely with our business partners. Agile is not just for teams; it is for everyone, and it's how we transform our entire organization, not just part of it. Wow.

Rik also acknowledged the business partners in the room, and challenged them to bring others in their organization along on the journey. Don't leave it all up to tech to get the right people in the room, and don't wait for it to be perfect. Ask questions. When in doubt about attending a ceremony or planning meeting, call the organizer or someone on the team and ask them about the intent and outcomes of it.

"Don't ask if you should attend. That is up to you. Ask about intent and outcomes. When you get your answer, you will be able to make a decision. And if you still can't make a decision, just attend with an open mind. We may not be as organized as we should be, but talking and working together, we can close some of the gaps."

Rick believes alignment is crucial to WL keeping pace with technology disruptions and rapid change. Our teams can be the best teams in the world, but without alignment, we are never going to be better than today. Rick warns not to mistake alignment for control. We are moving to have more decisions made where the work happens: with the teams. Alignment enables leaders to stop being bottlenecks, and teams to make local decisions.

Even though Rick nearly ran me and Vijay off the road about our cyber security problem, he is sounding good today. For every moment Vijay and I can encourage and guide Rick, there are probably 20 other moments a day where this isn't happening. Rick is going to be a great help to us, provided the pressure above him doesn't clamp down any tighter than it has already.

"Our first step in alignment is in this room right now," Rick gestures to the vision. "Talking with this team who gathered feedback, and harvested the opportunities and needs of WL. They collaborated to create a vision, which we will support."

"Thank you, Rick." Vijay is obviously pleasantly surprised by Rick's voice of support for their work.

"Of course. In past years, we would have beaten this vision to death for months before arriving at one that a committee would have developed. Today, we are witnessing history at WL, as we embrace a transformation vision that was created with a lightweight process."

Vijay jumps in, "So let's *live into* this vision by using it, and not changing it to meet our own needs. Instead, find ways to connect your work to it."

Rick nods, "Team, we leaders will give you our support. I will make sure of it. But I will need your help."

The room is stunned for a moment, in a good way. In his own style, Rick put his stamp of approval on our vision work, and basically said those in the room would support it. Don't try to pick it apart and change it. The business partners are nodding their heads in agreement, and the transformation team is giving each other a thumbs up.

Maybe Rick was a little too directive? I'm not sure. It's a nuance I only noticed because of my work with Eve. I will have to ask her about it. Actually, I don't think I need to ask her about this one. He spoke on behalf of the room, and that's never a good idea. His heart was in the right place, but it came out as a directive. He should have offered his support, and then invited others in the room to find a way to connect to the work. Discover it, challenge it, and hopefully, support it.

"Is there a deadline to adoption? Do you have a road map?" Thad asks.

Vijay shares the team's roadmap with us. We are used to having a detailed plan to march a project down the field. This is a much lighter tool. It still shows where we're going, and highlights milestones along the way, but it's only certain for the next six months.

"Have you considered a TPOC?" Thad is still on his laptop. I cannot believe this guy. He doesn't give lean a bad name; he gives VPs a bad name.

"A what?" Rick asks."

I try to stop the conversation, "Hold on— "

"A transformational plan of care," Vijay answers before Thad can. "I believe it's a lean method of …"

"Oh, yeah, we are working through that," Karen says while looking at me.

"It would behoove you to have this before you launch your vision." Thad warns. "I know you are excited to get going, but I feel you are not ready."

Ugh.

"What's a TPOC?" A business partner asks.

Thad ignores the question. "This launch seems premature. Perhaps I could offer some additional feedback after your report out."

"Well, this report out is designed for feedback," Vijay says. "I invite you to share it now. We want to encourage transparency for all involved."

Thad clears his throat, "I sense there is a need to better organize your transformation. A TPOC might be in order. There will be more rigor, and that will help you develop sound metrics for progress."

"What's a TPOC?" The business partner persists, now clearly irritated for being ignored.

I hold up a hand and ask everyone to stop the discussion. "This is a moment we have been waiting for, team. We have a business partner in the room with us asking for clarity. We can continue to talk over them, and anyone else who doesn't know what a TPOC is, or, we can answer the question. One conversation at a time, we have a choice to build trust or erode it, to draw people in to our work, or turn them away. What do you say?"

The business partner smiles and offers a quiet thank you.

"A TPOC is a transformational plan of care," Vijay says.

Thad jumps in before Vijay has a chance to finish. "Our team has one. It's a plan that describes how an organization will prepare and transform itself to maximize the potential of its investment in the change, in order to achieve the identified outcomes. Don't get too wound up on the acronym. It's just a tool to make sure you invest in and execute the change you committed to, and have a way to measure it."

First, I'm impressed that Vijay knows what a TPOC is. I didn't. Then, Thad…his team is doing one? And, for a guy who didn't show up at the last report out, and arrived late to this one, he sure has a lot to say now.

I have the floor, and pull the group back to Thad's question about a deadline and roadmap. Our transformation to agile, lean, and any other continuous improvement methodology is a different kind of change.

Everyone has their own pace of adopting change, so we can't force them to do it on a schedule. If we pile on too much at once, and too fast, we will only create chaos.

I've read about many companies that have *gone agile* before WL, and messed it up. They approached their transformation as installing agile, with a prescribed set of activities for people to become agile, like training and *running scrum*. But what ends up happening is that people keep doing their work relatively the same, but they call it something different. So, we are working to understand the current state of WL and adjust our approach accordingly. It might sound strange, but our transformation is so much less about methodology than it is about helping our teams and leaders be successful. I didn't understand this idea at first, because I thought agile and lean were tools that we need to learn to use to get better at tech. Methodologies have tools, but they are mainly mindset changes.

The way we currently run projects isn't working well. The way we currently engage our customers and business partners isn't working well. We don't have bad people. We have bad processes and an environment that doesn't allow for us to free up our time and energy to focus on what matters most. Changing this for even just one team is a colossal event that will take time, and us relentlessly sharing our vision of our future state. Not a final arrival, a future.

Right or wrong, the decision to *go agile* was already made for WL. We can and must tailor how we adopt it to WL, at first by using our conversations and survey data. We value strategy and roadmaps to plan our work, but not at the expense of our teams, the people who actually deliver customer value.

The team has learned from others in the industry that agile is a catalyst for organizational change, not a final destination. When we see agile as a catalyst, we can focus on leading the change with organizational change approaches, and what the agile community has successfully used.

I surprise myself with my answers. But when I look at Karen and Vijay, they are not at all alarmed. They are smiling at me.

"What if a leader doesn't want their team to go agile?" Vladimir asks.

We have an important strategy for resistance that we invite this group to review, reflect on, and then figure out how to use your style to lead with it. We strongly believe our teams and leaders are willing to change, but they need to be mentally equipped and supported to do so. They also need to understand the compelling reason WL needs to *be* agile, not just told *we are going agile.*

When people ask us questions and challenge us, we will embrace it, and see it as opportunity. People who want to learn more show that they care about what the change means to them. They are hungry for connection. We can then offer a custom conversation of leading change, which might include empathy or success stories showing the value of change. These are the moments that are not about agile, but about how to help people be successful. We need your help doing this over and over, for a very long time.

Since people process change differently, our work plans are definitely going to have to adjust with what we learn. One example of how we adjust is in our conversations with people. I can't expect to use the same conversation about the value of adopting agile with our CIO as with one of our business partners. The vision is the same, but explaining the business value, the compelling reasons to change, needs to be adjusted.

I might have gone a little deep with this topic, but this group was asking all kinds of questions. They seemed ripe for the discussion. I hope I picked the right group, and said the right things. If not, I guess I can adjust it for the next report out.

I offer the floor back to Karen. She shares more of what the team has planned, most if it is around getting the right people together to get off to a good start. Someone wants to know *who will be the right people?* Karen said the team doesn't have a firm answer on that, because they are still learning what the next step of this change is.

Someone else asks, how they could come up with a vision, if they don't know who is going to carry it out? Karen affirms that they are the vision team, not necessarily the transformation team. It's a foregone conclusion that many of the people on this team will be on the transformation team, but no assumptions are made. Carrying out the transformation may need different skills and talents. This makes some people uncomfortable, including Vladimir.

Ben senses this discomfort, and offers encouragement. Our old way of having the entire plan laid out ahead of time may have felt more comfortable, but in the long run, it was a rigid approach that didn't allow for learning along the way. We made many changes, and when we did, they often felt like failure. Then, if we did learn something and had to pivot, usually a person or a team was blamed for the change. We are taking this a step at a time, so that we can allow for discovery and learning along the way. This allows us to adjust without major rework. We will make all of our work visible, so anyone can see our progress, at any time.

Ben asks if anyone from the new scrum teams has feedback for the group. One of the team members said that since he began working in this new way, he couldn't imagine ever working the old way again. He gave it a chance, and it's different, and fantastic all at once. He said the last few weeks have been the best weeks of his career so far, and he knows it can get even better. He never imagined a change could be this dramatic, but it is.

"Does that mean your transformation team will run like a scrum team?" Vladimir asks.

Vijay wisely tells Vladimir that their first priority was the vision, and next will be to get the right people on the team. Back to the point about this transformation not being a calculated roll-out, the team doesn't want to form as a scrum team because it looks good. If they find value in going scrum, they will try it.

The only things we have for certain are a vision, and a plan to work in an iterative way. "Seeing our transformation in increments will allow us to adjust and adapt as we go. Our vision will ground us through it all." Vijay has a calm way about him that helps the room relax.

"It's going to feel different. But with our vision, your support and leadership, we will be successful." Karen offers. The leaders in the room nod in agreement. At least, that's what it looks like they are doing. For all I know, they could be nodding in agreement that we are all nuts.

Karen has done such fantastic work with Vijay. She doesn't report to me, and has her own responsibilities outside of the transformation vision work. She doesn't have to be here, and she doesn't have to care this much.

Vijay reports to me, so he's had the luxury of offloading some responsibilities to others. Karen is the best volunteer I could hope for. It's clear she's passionate about change and about agile. I will make sure to give her and the team feedback after this event. They have far exceeded my expectations of helping others understand what we are doing…when we barely know what we are doing.

Karen asks for any other feedback. If there is no other feedback, we will move on to the naming of our work.

Vlad asks how we are going to get the right people in the room. What if we miss someone? With WL so large, how will we know? Vijay tells Vlad that there is no guarantee. We have a fairly good view of the organization, and we can do our best. If we miss someone or some group, we can adjust and keep moving forward. Vijay reminds everyone that this is not the only time we will do something like this. Each time we work this way, we can sense, adjust and adapt.

"You mean *plan, do, check, adjust*," Thad corrects Vijay. "PDCA."

Vijay nods and just agrees with Thad. Huh. That worked. The group didn't clamor around the words, and Thad let it go. It's not a competition, but score one for Vijay.

The handful of business partners in the room are very excited about what they heard. They volunteered their time to be part of the transformation team without even verifying with their director first. They said their director would absolutely buy into this work, and if she didn't, they would bring her along to a sprint review to see it herself. Karen challenged the business partners to do that anyway. The more people see the value of what we are doing, the easier it will be for them to connect it to their work.

The business partners' excitement felt great. In all the uncertain work of this transformation vision, seeing them so engaged and willing to help validated our work. It felt worth it. We aren't doing this work to please them, but we desperately need them for our success. Speaking of success, they also seemed to understand that we won't always get it right, and that we'll fail, and have to try again. Hearing them say the word *iterate* was an incredible moment in my career.

And we were hesitant to invite them. We were worried there wouldn't be enough detail for them to understand what we were trying to do. Normally, we provide them with fully baked plans. Even grounded in the agile manifesto, our work could appear reckless and scare them away. They seem to get it, and want to learn more about it.

We're onto the unveiling of our new transformation name. It's one of the most exciting moments for this team and for me. The current name is Project Learn. It even sounds like a corporate initiative. Aside from sounding corporate-y, the *Learn* part of it is easily confused with Lean. Given that we were forbidden by the company to say the word lean, this name was ripe for rebranding.

The team volunteered me to report out on this work. I wish Eve was here, because this moment feels big in my journey to become a better leader, and to understand agile.

"Much of our research showed that we don't need to name our transformation. If we are changing the way we are working, why name it? It's just…the way we work. Our reality challenges this thought, because this is a very big change for WL. We feel people need a way to label the transformation, to name themselves while they are…becoming."

Becoming is the name for our transformation. It fits very well with our vision: *We are continuously learning how to better create and deliver ways for our customers to connect to the world.* We are becoming more value focused, becoming more customer focused, becoming a continuous learning organization, becoming an organization that can *grow happy* and *grow strong*.

Becoming is a very agile word, because it gives us the idea of increments or iterations. We aren't going to *go agile* overnight or even in two years. But, day by day, we will become agile. We won't apply lean overnight; we will adopt lean principles over time. And, as we work toward becoming, we will change ourselves along the way. We will learn what works for us, and what doesn't. We will sense and adapt our becoming.

Becoming is also a personal word. Because mindset changes like those of agile, lean, and DevOps are absorbed by people in different ways, it can't be rolled out or forced. This is true of leaders and team members.

"Speaking of truth," I say, "I want to be very clear that we intend to be as transparent as possible with our vision. We will use industry standard words such as lean. We understand this will take some collaboration with our HR and communication partners, and we are willing to do it."

"What are you going to say if someone asks you if there will be layoffs with our lean transformation?" Thad asks. It's as if he already has plans for them. He probably does have plans for them.

"If someone asked me today, I would tell them we have no strategy for our lean transformation yet. I would also say that no jobs at WL are ever guaranteed."

"Well, of course. But in the future, you will get that question, Joel." Thad warns.

"Do you know something about our lean transformation that this transformation team doesn't know? If you do, now is a great time to share."

Thad chuckles, "No, of course I don't. I just have experience with this sort of thing, and lean transformations are notorious for layoffs. People will be onto that. You should keep that top of mind."

"Understood," I nod, trying to keep my posture relaxed.

"I don't see anyone from corporate communications here today. Did you invite them?"

"Yes, Thad. HR was invited as well. Neither are represented today, so the transformation team, once we have one, will follow up with them."

I expected Rick to challenge our decision to use the word lean, but he said nothing. I guess if the lean expert is okay with it, then he should be too. It's probably not that simple. Maybe he is waiting for hell to break lose when we talk with our HR and corporate communications stakeholders.

Thad asks one more question: how will we measure success?

My turn to rest. Vijay assures Thad that when the transformation team forms around this vision, they will develop metrics so we can track progress. Our metrics will change over time; what makes sense on day one of our transformation probably won't make sense one year later.

Thad urges Vijay to *draw a line in the sand* now, before the work gets away from us. This is big baseball, and as such, needs the attention of such a grand thing. Vijay takes the baseball comment in stride, and assures Thad they will have metrics. Thad nods, and then folds his laptop, and leaves the room.

The report out is concluded, and everyone high fives one another. Even Rick gets in on the celebration. We're exhausted and optimistic.

Reflections

People working with me have passion and energy for change.

Our business partners are more ready than we thought.

I have a good idea of who could be a successful part of the transformation team.

It's a relief to speak out about using the word lean.

A methodology war is brewing.

Executive Whimsy

Walking to J&L's Café for my coaching session with Eve, I'm feeling conflicted. That's nothing new, I guess.

As we settle ourselves at a table, Eve pulls out my vision that we built together, just like she does at the start of all our discussions.

I see the first priority in my vision, *To be trusted by Lora and Rick*, and point to it, "Looks like we should adjust my vision, because Lora's not even at WL anymore."

Eve nods, "Hopefully, there is more to adjust than just removing Lora's name. But I want you to have time to reflect on that first, so let's save that for another assignment."

"You're letting me off the hook on something?" I chide. I know better.

"Never," Eve smiles. "What's on your mind today?"

"Too much," I sigh.

"Did you lay out the path to the transformation, like you did for me?"

I laugh, "I did, and even had a roadmap. But it took a few turns."

"You look conflicted."

"We changed the name of the transformation from *Project Learn* to *Becoming*. Now I'm not sure we should have named it at all."

Eve sits back in her chair, but says nothing. She's expecting more from me. More context, more thoughts, and more problem solving. Our coaching relationship isn't about me coming to her with the problem of the week, and then getting direction on what I should do so I can turn around and do it. Our interactions are much more about mindset: how am I thinking about an opportunity or problem? Then, Eve uses concepts from methodologies like agile and lean to offer guidance to help build my mindset.

"The team worked together to come up with a name for the transformation. They were so jazzed about changing the name, joking that just about anything would be better than Project Learn. They even considered not naming it. So we really explored all the options we saw available to us at that time."

"I remember that, Joel. *Project Lean.*"

"Right."

"You don't have to name a transformation. You just have to visualize it, communicate it, and lead it."

I sigh, "I agree. It sounds so simple when you say it."

Eve shrugs.

"So that's where I'm hung up, Eve. The name *Becoming*. It just doesn't sound good to me now. I feel like we should just not have a name. We can talk about transforming technology without naming it."

"You don't like how the name sounds, or you don't like that you named the transformation at all?"

"I think the name itself. It feels too abstract. This change is for the masses. Although WL has a bunch of super smart people…"

"Did you tell your team how you felt?"

"No. We already did all of the work, and agreed on it. I can't come in now and change it."

"So, you don't like the name, and, you don't want to tell the team that you don't like the name," Eve pauses. "If you *did* tell them, would you want them to come up with a new name?"

"Uh…Yeah, that's the thing," I'm uncomfortable. I know I'm wrong; I need help getting out of this mental spot. At least I recognize it! "I don't have a better name. And yet there is the idea of just not having a name, but that's only because I don't have a *better* name."

Eve smiles. Waits.

"What do you think of the name? You've probably seen a bunch of transformations."

Eve sips her matcha latte, "Doesn't matter what I think."

"Well, if it was awful, I'm sure you would have told me."

"Why would you want my opinion on a name that you don't like?" Eve asks.

I chuckle, "Got me."

Eve's still smiling, "You're catching yourself, Joel. It's great to see you grow."

"Well it doesn't feel very good."

"No," she leans forward, "but no one said it would."

I sigh.

"Joel, you already said what should or shouldn't happen. You know how to lead for this situation. Just slow down enough to see it."

Great, I already said the answer and don't know it. What could I have said that was the way through this? I know I can't stop the team because I personally don't like the name, so…? Ugh. I put my head down on the café table, and close my eyes.

She says I already know the answer, and even said it. All I said was that the team already decided this naming thing, and that I can't really go back to them to change it because I personally don't like it. *Personally don't like it*…ugh. The answer is clear: the team decision should stand.

I pick up my head and smile at Eve, "I *do* know the answer. I can't override the team on a team decision. Not just because I have second thoughts about it later on."

"What is that, Joel?"

"It's not really *trust the teams*…And it's not really *what* versus *how*…"

Eve waits. Again.

"I think the report out just messed with my head."

"The report out, or a specific person?"

I laugh at her statement, "You know me better than me sometimes. A specific person. Thad."

"I just see it from the outside perspective," Eve says.

"So I just had my foundation shaken a little, and when that happened, *Becoming* suddenly sounded out of place."

"Back to the concept or methodology where this belongs. There is something called a silver bullet."

"I've heard of that."

"An executive override. Used very sparingly. Maybe once or twice a year. Usually due to extenuating circumstances that are hopefully based on newly learned data, or newfound funds to invest in something no one thought they had money for."

"But not for personal reasons," I lean back in my chair.

"Right. That's executive whimsy," Eve smiles. "You did great with this, Joel. You realized you were trying to make a decision based on emotion, and so you stepped back. Plenty of leaders don't pause, and follow through with their wishes, creating chaos, doubt, even lack of trust."

I tell Eve I have lived through this many times. I have probably done this to teams many times, too. She assures me WL is no different than other companies. The best thing to do is to forget about the past.

"About the interactions in the report out. I'm not going to lecture you about how you're just beginning to feel the heat of leading a transformation."

I nod, "Thanks. This was different, Eve. Thad is a very challenging person. He just got a little too far into my head, and I lost my confidence."

"No, you let him too far into your head and he walked up and stole your confidence."

Boom.

Our time is just about over. I just shake my head and sigh, "I came to this meeting to talk about something I thought was very important, and yet we ended up talking about the most important thing."

Eve smiles, "You're getting better at reflection, Joel. It will help you exponentially."

"Just when I think I am looking at a problem from far enough away, you show me how to step back even further."

"Almost always, there is room to give yourself another step back."

Reflections

How to keep Thad from getting so far in my head?

How to do a better job protecting my confidence?

When I reflect, there is always room to step back a little further.

Reflection feels like the best way to keep Thad out of my head.

C²   .

It's the inaugural meeting of the newly formed C² committee and I am a part of it. C² is short for the Culture Committee. Fresh out of the vision event, culture is the hot, new item that has the executives' attention, and so a steering committee was formed. And because Thad's words about having a *stiff head-wind of culture* were discussed in our team's breakout session, the group thought it would be good to ask him to lead the charge. It's ironic that he's been selected, because in the short time he's been a VP here, he's one of the most command-and-control leaders I've experienced. He says all the right words about how he wants to help WL change for the better, and he's new, so he's perfect committee leader material.

Jack and Meryll were giddy with excitement that I was on the committee. They didn't have to say *careful what you wish for*, but they did. Aside from my membership being a source of entertainment for my two favorite peers, there is likely a connection to the transformation work I've been asked to lead.

Dipti is listed as a co-leader of the committee. Actually, in an email, Thad wrote that he is *spearheading* the committee and Dipti is his *second chair*. But Thad couldn't make it to this meeting due to a more pressing issue, so Dipti is ready to lead us. That itself is a cultural issue: leaders showing up. Prioritizing.

It's a big group, even by our company's standards of doing things by (often belabored) consensus-style decisions. Several managers and directors from all over the enterprise are present in the room, and on the conference call. Technology is represented in the room by me, Thad, Dipti, and Gabriel. We were selected by Thad and Rick to join the committee because technology needed strong leadership representation and voices of change. With Thad not present at the highly visible kickoff meeting, I'm convinced we were selected to be his backups. It doesn't matter now, I guess. I'm here, and I'm very curious about what a culture committee can accomplish. And, I respect Dipti and want to help her.

We are welcomed to the committee, and told this is a very exciting time for us, because we are going to change ourselves for the better. Realizing our vision to *grow happy, grow strong* is dependent upon transforming our culture, and that is why we are here. Then, we are asked to go around the room and say our name, and what area we represent. That takes almost 15 minutes of our 90-minute meeting, because some people offer more than their name and area; they share their excitement about changing the culture. Some of them must think Dipti is at Thad's level, and need to posture themselves right out of the gate. It's classic steering committee behavior.

"Our first order of business is to get ideas on how to change the culture in our enterprise. The goal is to brainstorm a big list of options to help us get started. Skype people can use messaging to send in your ideas. Someone in the group here will capture them and add them to the list."

"Can I ask a question?" A manager raises her hand, and Dipti encourages her to ask it.

"I don't mean to be disrespectful, but what is the direction of this committee? What have we been charged to actually *do*?"

Dipti smiles, "Well, it's to help realize the vision to *grow happy, grow strong*."

"Yes, but why is culture going to help realize the vision? I don't mean to be a pain," the manager shrugs.

A director chimes in, "I think what she's trying to ask is, does C² have a vision? Do we have something we can ground our work in?"

Someone else comments about the C² name; how can we name a group that has no purpose or vision? And, isn't *C3* an explosive? And why would we name something so close to the name of something that explodes? The comments are loud enough for many of us to hear, but quiet enough for us to ignore it. With or without Eve coaching me, I know this is passive-aggressive behavior and poor leadership.

But…over and over, we are asked to do things like this, and then told to *drive for results. Deliver.* This environment has turned even the most positive thinking people into cynics. I'm sure there are many outcomes from this environment; I only know the ones I have seen and felt. My experience is that when people are put in uncomfortable positions like this, they say and do strange and usually negative things.

I don't think I should speak up about it right now, and take this entire meeting off the rails. And then there is the risk of this being a cross-functional committee. The last thing I want to happen is for Rick or Thad to hear random feedback from someone in Controllers that I am rocking the boat or holding the team back with my negativity. I want to be a change agent, but I also want to keep my job.

Dipti shifts in her seat, "Thad isn't here, but I think he wants this committee to come up with ways to change the culture any way we can. I think we have a blank canvas. Well, as long as it leads to helping the enterprise to *grow happy, grow strong.*"

Now a conversation about what Thad might want for the group ensues. We are speculating at best, and everyone knows it. We are used to working this way; trying to do work without vision or without the visionary leader in the room. We are also used to what happens later: the work the committee accomplished is often changed when the visionary leader or leaders learn about it. The decisions and ideas are swept aside or removed all together. Usually we are then asked to begin again, this time with a lot less motivation and energy. This is when the laptops come out and committee members begin multitasking during the meeting, or skip the meeting altogether. It doesn't do any good to get frustrated or voice concerns about it, because then you will be labeled as someone who doesn't want to collaborate with others. Yet working on something that doesn't have a foundation doesn't help anyone *grow happy* or *grow strong.*

Yes, this discussion is heading that way. After a good 15 minutes of speculating on what Thad and the company wants C² to do, we decide to proceed with the brainstorming activity. No one wants to be a jerk about it to Dipti, because we all know she is a messenger. She's fully capable of creating a vision with us, and leading us with it, but everyone knows that's not what she's expected to do. We've all experienced this before, and are happy it's not us in the hot seat…this time. Dipti looks relieved to get the brainstorming activity started.

We generate a long list of things we can do to change the culture:

A one-stop culture web page
Culture communities of practice for different parts of the enterprise
Culture team, to carry out the ideas of this team
Video series on how to work differently
Culture Talks, leader presentations to employees
Outside culture experts and speakers
Culture newsletter
Culture Ambassadors
Culture fair
Hire outside culture consultant
CEO messages about culture
Culture buttons, mini rewards for employees who live the change
Culture games throughout the company
Culture T-shirts
Culture Bux, for employees who live the change
Culture signs in walkways
Culture PowerPoint template

This group really got into brainstorming, and we delivered a robust brainstorm list. Most of the ideas have been recycled from other efforts, but it felt like a good start. Dipti was a great facilitator, stopping any comments that dug more into the details of each idea. So, when someone would ask the person who gave the idea "What would that look like?" or "I can't see what that is," she would gently but firmly steer them back to brainstorming mode. Where anything is possible. Even though all of us know better.

In the final part of the meeting, we are asked to vote on the top five activities. There is something not sitting right with me at this point. My gut is telling me that we are doing this in a vacuum, even though we are a very large, cross-functional committee.

"One more thing before we vote," I say as gently as I can. "Might we consider getting feedback from our employees on this? I'm no expert, but it feels like we are missing a big part of what this could be." I don't even mention the part about how this list feels like only the employees need to change, that the leaders are handing it down to them. Or worse yet, forcing it on them. It's not like me to hold back, especially with Eve coaching me, but this is a big enterprise committee in its first meeting; I can't alienate people on the first day. These people are not likely getting any agile or lean leader coaching, so I could be too disruptive.

"What would that look like?" Another director asks. "I'm not sure I can see how employees can weigh in on their own culture."

"How would we ask them?"

"What if we survey them, and only get a low response rate? Does that count?"

"What if our competition gets wind of it and uses it to recruit people away from us?"

"What if we survey them and their thoughts are counter to *grow happy, grow strong*?"

"Why aren't there capital letters used in our vision?"

And on it went for several minutes, until I was sorry I asked, and much of the group was wandering in the wilderness of *what-if*. Again. The questions were really objections in disguise. No one wants to outright shoot down an idea on the first day of the committee, just like me not wanting to speak up. So we have an elaborate dance around reality without ever embracing it.

The room is then silent. Dipti patiently waits for more questions.

"Are we ready to vote?" she asks. A few nods are enough to give the committee the go-ahead.

We vote on the top six things from our list, and then divided into teams to begin working on them. The ideas that made the cut are:

Culture Web Page
Culture Games
Culture Bux and some kind of reward system
Culture Newsletter
Culture Fair
Culture Ambassadors

A few C² members were disappointed that hiring an outside consulting firm didn't make the list. They were already talking among themselves about how to cobble together money from their budgets to hire one anyway, just for their areas, as a test. Classic WL culture, to break off and do your own thing, so you can continue to feed your favorite work, regardless of company strategy or in this case, a group vote.

The group also agreed that C² would be a nice name for the overall effort, only in this context, it means culture change. So now our catchy name has two meanings. I hope we can keep it straight as we progress forward.

"I know most of the group agreed on this, but the more I think about it…" Gabriel hesitates. "Won't it be confusing to have one name for two different things?"

Well, that's a relief. Meetings like this often cause me to doubt myself more than I should. It's the enterprise view that is a challenge. Thinking about the entire organization, instead of just my world in WL.

"What do you mean?" Another manager asks but probably doesn't care. They really are saying, *Hey, I thought we agreed on this and are ready to move on. You're messing up our progress!*

"Well, I mean exactly that." Gabriel uses caution. "Won't it be confusing to people that we have a committee and an overall effort by the same name?"

The room talks about it for 20 minutes. We arrive at the same decision, agreeing that no one will really know about our committee, so it's acceptable to have the same name. At the start of this meeting, I wasn't sure if there were too many people on this committee. After the dragged-out decision we just made, I'm convinced of it.

"Do we have a budget?" A manager asks. "Because I'm not sure how we can get started on these things if I don't have a number to charge them to."

Dipti admits there is a budget for this work, because it has the attention and backing of our executive leaders, but she is not sure how much we have to work with, or what the number is. Thad was going to get that information to her, but it didn't happen.

Someone started pressing Dipti for more information, so she had to offer that Thad had canceled the last two meetings they had together because of other priorities. She hasn't had a chance to catch up with him. Now Dipti was uncomfortable; the manager who questioned her apologized. Dipti said it was no big deal.

Then someone asked about the time commitment for this committee. Was there a sense for how much effort is needed? The ideas we generated were great; but who is going to carry them forward?

"Well, we are." Dipti smiles. "We have a very talented group of cross-functional leaders."

And, a few secret budget dollars.

Then someone asked for an estimate of the percentage of time we would have to commit to the effort. This is a nice deflection from owning the fact that we will all have to do actual work, not just talk about it.

Dipti said she couldn't answer that yet, because we don't have a budget. She assured us that she would raise our questions to Thad, when she sees him. Thad is definitely a bottleneck for this work.

One manager was concerned. "I'm confused. Why did we just crank out this big, awesome list of ideas, if we don't have any sense for how committed we are to the effort?"

He had a great point. We don't have a vision or a visionary, we don't have a budget, and we don't have any idea how much time the work is going to take. Yet we just generated a list of work that could last for years, if we wanted it to. Needed it to.

Dipti sighed, apologized again, and repeated that she would take our concerns to Thad. Hopefully, Thad could attend the next meeting and have answers for us.

The manager apologized for putting Dipti on the spot, but surely she understood how we are all pulled in multiple directions. Things are moving so fast, and we have to put our efforts where they can best be leveraged. It will be difficult to commit to such an ambiguous effort.

I tried to picture Eve in this meeting. She is so good at seeing beyond the obvious corporate function or dysfunction to reveal a root cause. What would she say or do here? The bottleneck issue would surely get her attention. She would look at both sides of it…

Bingo. Two sided. This bottleneck has two parts to it. The leader *and* the teams. Probably most bottlenecks are like this. In this case, Thad is causing the bottleneck, but we are allowing it to happen. Are we forced to allow it, or do we only perceive we are?

We could absolutely determine our vision based on the ideas we voted on. We know what goes into that work and have lived through it many times. We might spin for a meeting or two, but we would probably spend far less time doing that than what we will spend waiting for Thad.

We could establish a budget based on the ideas we voted on. We are all leaders familiar with the budgeting process. With so many in the room, that might take a while, too, but no one is going to let the group be reckless. We know we have money, or a committee would have never been formed including so many different parts of the company. We could really capitalize on that part of it, because everyone loves a good 'ole fashioned cross-functional effort. It makes us feel like one company after the same goals.

Once we have a vision and a budget, we could determine our own commitment to the work, which would give us a way to prioritize what gets done first. All of this could be presented to Thad and the stakeholders for buy in. They can give us feedback, and we could adjust as necessary.

Just thinking of it gives me a rush of energy. Imagine if we actually did the things we were hired to do, without reservation or fear of retaliation. We would be great. Might be a little challenging to get started, because we are so accustomed to everything being decided for us or ahead of us, but then we'd be great. Not everyone would like working this way, but I imagine after a few brave people step forward, don't get fired for it, and actually get praise for it, more people would want to get involved.

But I can't offer this. I would be viewed as a complete radical. No one would follow me on it. I'm not feeling like I want to lead the charge on this alone. Especially with Thad as the leader of the committee. Eve will tell me that's exactly why I *should* speak up, but I just can't do it. With the bad news about our competitor launching Green, and the regulatory fine for our cyber security problems, people are even less likely to take a chance than before. No one wants to stick out from the crowd right now. I guess I don't, either.

Dipti assured him that she felt exactly the same way. Then, she sincerely thanked all of us for our good work, and dismissed the group, before anyone else asked her another question that she didn't know how to answer. A wise move that made me respect her more.

Given that Thad is a bottleneck to my work with the agile transformation, and with C^2, maybe Dipti and I will have a need to work together to get what we need.

Reflections

Culture is a hot topic across the company.

Vision is desperately needed to make decisions, and get buy in.

How much of our bottlenecks are the leader's fault vs. the team's fault?

I see more things that need to change, but they are not worth bringing forward. This is focus.

Second Sprint Review

"Hi. I'm Waylon, scrum master for *Can't Make This Up* and host of this sprint review. Welcome to the second ever sprint review at WL. Someday soon, we can stop counting them, but for us, it's still a pretty big deal."

Waylon shared the sprint goals of the three teams, and stated which ones were met, and which ones were not met.

I'm sitting with Meryll, Jack, and a few other peers. Even Karen has found her way into this event. Of course she has. She can't wait for WL to change for the better. When we entered the room, we received a handout, to help educate us on the ceremony.

Welcome to WL's Second Sprint Review

Teams:

Clever Hashtag

Can't Make This Up

It's Never Like This (INLT)

Why do agile teams have a sprint review?

To give our customers and stakeholders a chance to see working features and provide feedback.

To give our teams a chance to show the contributions they have made to help solve the business problem. We can also take some pride in our work and progress.

To give closure to two weeks of hard work. Many people contributed to provide new value to our business clients, so it's time to show it.

Thank you for coming today. We look forward to your feedback.

There were more people at this review than the first one, but it was still thinly attended. The team did a great job of promoting it through email, but there are probably more places to get attention for this event than email. Karen brought friends and coworkers, which provided some new, curious faces.

I loved seeing Vlad walk in. For the first sprint review, he really struggled to understand what was going on, including why team members were wearing jeans. It's clear now that he's more curious than anything. I'm so glad he came back. He sits down near us, and we smile at him.

"We're going to show you what we got done, here, but the most important reason we have this event is to have in-depth conversation and collaboration with all of you. So, this event isn't just a one-way flow of information from our teams to you. I invite you to think of it as a scheduled opportunity to inspect and adapt our work."

"He's good," Vlad leans over to us. We nod in impressed agreement.

Just then, I see Thad standing in the doorway of the room, lingering. He's taking a bunch of notes in a small notebook. Huh. It's great that he's here. Wonder why all the copious notes?

The mood was positive and energetic, until the team members tried to show us their work…and couldn't. At least not without a long wait. The test environment was very slow. They apologized over and over for the system taking so long. The silence in the room was thick. They were all so uncomfortable, that most of us felt it, and we also became uncomfortable. A strange moment. Yet the teams persevered.

Waylon and the other scrum masters were helping out on the sidelines, but they didn't take over. Neither did the three teams' product owners, the people who represent the business and voice of the customer, and are responsible for working with the business/customer to determine what work the team does when.

Those of us who are aware of WL's incredibly slow test environment felt the teams' pain. And, it didn't seem right that they were apologizing for it. The software they created did not cause the slow-down. The slow system holds back what they created. The slow system is an impediment.

The painful moments are over, and we are on to the question and answer part of the sprint review. There was a really good question from one of Karen's coworkers; a functionality that was needed, that the team hadn't thought of. It was a bright moment of collaboration that everyone could witness, without any additional explanation needed. While this happened, Thad was still taking notes in the back.

Now we are standing around, talking with the product owners and scrum team members. Ben walks up to us and apologizes for the slow demo.

"That's okay, Ben." Meryll touches his sleeve. "We know how crappy and slow our test environment can be."

"It's an impediment, not an apology," I say.

Ben asks me to repeat what I said, so I do. From the look on his face, he must have originally thought I was complaining, or being sarcastic. He contemplates my words, then slowly smiles.

"Yep. Waylon needs to hear that, Joel. So he can share it with the other scrum masters, and the teams. They are pretty beat down right now. They powered through it, but it was tough."

"But that impediment…we should do something about it," Vladimir says. I just love his enthusiasm.

"One in a long list that's on the backlog," Ben sighs. "We are making progress. It will be interesting how some of these bigger impediments are fixed. Or at least, mitigated."

"I would like to see the backlogs some time," Vlad says.

"Of course, Vlad. Let's set something up." Ben has a new leader in his corner; that must feel pretty good. The two of them pull out their phones, and begin collaborating on when they will get together.

Jack leans over to me and Meryll, "Thad's a lurker in the back."

"Yeah, he's taking a mess of notes, too."

"Maybe he wants to be a scrum master, or something," Jack laughs.

"Don't even…" Meryll mumbles.

"Ah well, he's gone anyway. Wouldn't want him to do the socially acceptable thing and mingle, to, you know, provide some feedback," Jack says.

"Jack, he can't make eye contact with us in a small meeting," Meryll says, "I can't picture him interacting in this type of event anyway."

I sigh, "Can we stop speculating about Thad? You two are letting him ruin the mood, and he's not even in the room. Let's have a good time, here."

"On a more positive note, we have more people at the event," Jack looks around the room. "Looks like Karen brought an entire posse with her. A good thing too, because that one dude knew all about the functionality the team was trying to accomplish. That was super cool to watch happen. It was the best advertising for agile that we could have hoped for."

"The incredible magic of making work visible, and then talking about it."

Reflections

Teams are resilient. How long will that last?

When the work is visible, it can be made better.

Teams want to make excellence, best quality.

Thad is way too much into our heads.

What do I Own?

I'm standing in line to order my coffee at J&L's Café. I have a coaching appointment with Eve in a few moments.

"Joel?" A strange voice is behind me.

I turn around to see Tony, a guy I used to play soccer and basketball with in high school. We called him The Mouth behind his back, because he had to dominate every conversation. He wasn't mean; he was annoying. No one knows more than he knows. No matter what you've done, Tony has always done it better or with more risk. One of those people.

He wasn't my friend and he wasn't my enemy. He was someone I put up with because he was a teammate. We didn't go to the same college, but our paths crossed at a few different running events throughout college. Tony always beat me. That didn't bother me, but how he bragged about his victory did. Then he would offer me all sorts of nutrition and training tips, even though I didn't ask for them. Other people thought it was hilarious. I couldn't stand Tony.

"Hey, Tony. How are you?" We shake hands.

"Fantastic, thanks. You're looking fit, you bastard. Are you still running?" Tony gives me a playful, harmless punch in the stomach. I'm 16 years old, all over again.

"Yes I am still running," I smile, but don't offer any other details. Tony looks pretty fit himself, but I'm not offering that, either.

Tony nods, as if he already knew, "Great. Great. We had some good times together in high school, eh? You know, I'm registered to race Ironman Boulder this year. It's only four weeks away. It's my second one. My wife is doing it too. She's gonna be great. You ever do an Ironman triathlon?"

Crap. Here we go. I have to race this fool? What are the odds? I start thinking about all of those times Tony beat me in college. Then I remembered he's a really good swimmer. These memories don't matter now, but I keep thinking of them. They must matter to me on some level. I could lie and say no, but then I would surely see him out on the race course.

"No, I haven't, but I'm registered to—"

Tony cuts me off, "Aw, Joel, you are missing out. The Ironman triathlon is a 2.4 mile swim, a 112 mile bike, and a 26.2 mile run, a marathon. Wait," Tony touches my arm. "Did you just say you were registered?"

"I started to say I was," I shrug. "I'm registered for Ironman Boulder, too. My first one."

Tony punches me again, this time in the arm, "You're *in*, man! Only four weeks away 'til go-time. You must be training your ass off. I know I am. You know if you ever want to talk training or nutrition, just call me." Tony fishes in his portfolio for a business card, and hands it to me.

"Thanks." I take his card and put it in my pocket. I'm not going to read the card right now, because I don't want to get involved in another conversation. Eve is probably already at our table. I don't want to miss a moment, because I have so much to discuss with her.

"Say, Joel, a few of us are heading out for a ride this week. Just a 30 miler for some hill training. Want to join us? Give me your card or your phone number and I'll text you the info."

I knew this was coming, and yet what could I do about it? I pull out a business card, and hand it to him. I know my wife Cele, who knows The Mouth, will say later, *Oh my gosh, Joel! Why did you give him your contact information?*

"Uh, I'm not sure, but let me know. I will see if I can make it."

Tony reads my card, "Director for WL, eh? I used to work for them. And I was a director a few years back. Still, your job is impressive, Joel!" Tony punches me one more time. I can hardly stand it.

My coffee is ready. Deliverance.

"I have to go." I shake Tony's hand. "Great to see you."

Tony has a broad grin. "You bet. My new Ironman Boulder bro."

I smile back. "Take care, Tony."

I don't know why The Mouth was in J&L's Café today, but I sure hope he's not starting a new habit. This place is not big enough for the two of us. Actually, I don't think the Ironman Boulder triathlon course is big enough for the two of us, either. It's a 2.4 mile swim, 112 miles of cycling, and a 26.2 mile run, a marathon. Ugh, those 140.6 miles seem impossibly long, until I imagine sharing the course with Tony. So much for being calm and focused before my race. In a few short minutes, this guy has found his way right back under my skin. Ugh.

I have to let this go. I can't let The Mouth ruin one of my most favorite parts of the week. I walk to my usual meeting spot with Eve, who is waiting there for me.

We start talking about C² and all the recent adventure I've had with it. Eve listens intently as I tell her about the senior leaders appointing the group, which is a cross functional team from all different areas of the company. I also tell her about our powerful brainstorming session.

"Let me guess, some of the ideas include culture ambassadors, a one-stop website on culture…" Eve pauses. "And, a culture fair."

Wow. She has her coach-poker face on, so I can't determine if she is joking with me. I know better. She's not joking, because she's in coach mode.

"Yes, as a matter of fact, we did have those ideas. And through voting, they bubbled up to the top as things we should do. Along with games, to get people engaged; a reward system to reinforce the behaviors called Culture Bux; and to help spread the change throughout the company, culture ambassadors."

We are sitting at a table inside J&L's Café, our usual meeting place. It's close to WL, they serve good coffee, and Eve likes their matcha lattes. It's a positive, upbeat venue, which is great, because often we are discussing difficult and uncomfortable topics. Eve has a way of pushing me like no one else. Getting shoved into agile coaching by Lora last year was one of the best and most challenging experiences I've ever had.

"Well that's a lot of ideas. At some point we still need to discuss your progress on leading the agile transformation at WL, and that side project of helping your CIO and his peers look for waste."

"Ugh, yes, all of that."

"No rush, Joel. I have it in my notes, so we won't let it fall. This C² work seems the most pressing. So let's explore it. Let go of the other things for now, as best you can."

I sigh, "Thank you for that."

We both take a pause, but I think this one is more for me than both of us. Eve is giving me a chance to refocus on only the C² discussion. Picture a coach letting the team rest after a set or a drill, to give the players a break physically and mentally. To reset, and begin again.

"What do you think, Joel?"

The break is over.

"Well, I wanted to ask *you* about it." I chuckle. Of *course* she wants my thoughts on it first.

Eve smiles, "I want to hear what you think, Joel. This is a really good topic."

"Easy for you to say."

"Only because I've seen it so many times before this one." Eve shrugs and sips her matcha latte. "Even the C^2 name."

I shake my head, "Ugh. We are a statistic!"

"It's okay, Joel. You're in a large, complex organization, and it's extremely challenging to even *talk* about making changes. Many companies don't even get this far."

I pause. I have been reflecting on this C^2 thing all week: I wrote about it in my journal, and thought about it when running or riding my bike. I have searched for why this work doesn't feel good. I have some ideas, but no huge revelations. I'm ready to share them with Eve. Of course she will reveal an entirely new perspective to me, part of the big value of having a coach.

"Your coaching has taught me to be curious first."

Eve raises her cup, "Love an idea for five minutes."

"Right. So I attended the meeting, listened to the conversation, considered the ideas. I reflected on the whole thing, and determined it doesn't feel right. And that I wanted to talk with you about it."

"I appreciate you sharing your thought process with me, and I love that you are practicing reflection, one of the five tenets of lean leadership. But you're not telling me anything, Joel," Eve says flatly. "What are you thinking? Try to keep the competition news and fines out of the equation, and just focus on C^2."

Once again, she's pushing me. I knew she would. I hate it and love it, all at once. I'm not dying from it, so I guess I can take another step.

I take a deep breath and start telling Eve that agile and lean were not designed to change culture. They are methodologies. Culture may change as an outcome.

The agile methodology is just one better way to develop software. There may be other ways discovered, but the people who got together to create agile and scrum believe their way is a better step in the right direction than the waterfall approach. Lean is a customer-focused methodology used to improve any process through continuous, incremental improvement, and removing waste. Both have mindsets associated with them. So, methodologies and mindsets change, and then culture change is an outcome.

Eve is impressed with my descriptions for agile and lean, and then presses me on culture being an outcome, "Does the methodology or mindset itself create the change?"

"No. The methodologies actually change how people work, and influence how people think about their work. But the actual environment…" I trail off. This discussion is interesting; I know I'm being led somewhere.

"What are you thinking, Joel?"

"We are arriving at some new place in my journey," I say slowly.

Satisfied, Eve sits back in her chair with a faint smile, and waits. She waits for me to discover something that has been right in front of me all along, but I didn't see it. Waits for me to realize that being an agile leader who also knows lean, is so much more than changing my mindset and changing how I work.

"Agile and lean change the environment of an organization. And when the environment changes, the culture changes," I say slowly. "Culture change is one outcome of agile and lean methodologies and thinking."

Eve continues to wait.

"Culture," I sigh, "is one example of many things that, over the years, our company has tried to change for the better, and have failed. People were blamed for the failure. But the people didn't fail. Well, mostly there are good-intentioned people who want to change things. The environment has failed the people."

I stop. I am jazzed and frustrated, all at once. I look at Eve who continues to wait with that faint smile of hers.

"Our senior leaders believe that our people hold back from changing the culture. If we can change the people, we can change the culture. Meanwhile, the environment stays the same."

"Joel, what do you mean by *environment*?"

"The way we fund projects. The fact that we *have* projects. The fact that we have so very many #1 priorities. The way HR evaluates performance. The way we are regulated. The way our organizations are set up to work together. The way we make decisions. The way we train our people. The way we get work done. The way we solve problems…" I put my head down on the table.

I pick up my head. "This feels similar to when I realized that our former VP Lora was crushing me with her leadership, and I in turn was crushing my teams with the same style of leadership."

"So Joel, you are saying that agile and lean methodologies and thinking change these things?"

"Yes," I am uncomfortable saying it. "At least, they can."

"You look so uncomfortable," Eve leans in. "What are you really thinking, Joel?"

I sigh, "This turns our discussion about me leading the agile and lean transformation at WL on its head."

Eve waits.

"Eve, that time we laid it all out, all the things I would do differently leading the transformation at WL, why didn't you stop me? Why didn't you tell me that those efforts would get us off to a great start, but would ultimately never work, unless we change the environment?"

"You mean, you didn't need a better name for your transformation than Project Learn?"

I shift in my seat, "Uh, I think we still need to do that."

"Your leaders don't need the same agile, and maybe even lean, training that your developers, your devs, are getting?"

"Yeah. We still need to do that."

Eve is steady, not scolding me. "Do you think you still should align with Thad? And relentlessly share your vision with all of WL? And get some coaches to support the newly agile teams?"

"Got it. I wasn't wrong. I need to do those things to lead the transformation," I sigh. "Well, we can, but it won't work."

"Then what?"

"I probably won't say this right…" I swirl the last of the medium roast in my cup. "We can't be successful with agile and lean if we just roll them out like we've done everything else. That's what we have been doing so far."

"What else?" Eve pushes me.

"We can't layer a transformation of any kind over a crummy environment," I sigh.

Eve encourages me to keep going.

"So, for a transformation to be successful, the environment has to change in two different ways. First, it has to change for the transformation to actually happen. And then, as a result of the transformation, the environment will be changed. And, if we are successful in changing the environment, one of the outcomes will be culture change."

Eve grins.

"That hurt," I say.

"You got it, Joel."

"Thanks, Eve."

"Why do you still have such a furrowed brow, Joel?"

"I think I'm just blown away. If I wouldn't have discovered this in time to lead the transformation at WL, would you have enlightened me?"

Eve's gaze is steady. "Has this coaching relationship failed you yet?"

"So the answer is no. You would have let me press forward, and probably learned it along the way."

"Has this relationship failed you yet?"

She's right. Again. "No."

It's silent for a moment. My brain needs a rest, and I'm sure I'm driving Eve nuts right now. Although she would never show it.

"I'm sorry," I say.

"It's okay, Joel," Eve smiles. "I'm so very proud of what you just learned. This was a huge win."

"Yes. And instead of worrying about why it took so long for me to learn this, I should just be happy I learned it."

"To learning," Eve holds her cup to me, and we toast.

I sigh and sit back in my chair for a rest. I am exhausted, but I have some other things to share with Eve. I hope she has time.

"Do you have more, or are you good for today?" Eve asks.

"I am full up, but I have one more thing to discuss. It's about bottlenecks."

"I love a good bottleneck challenge." Eve sits up straight in her chair. "But first, I want to give you your assignment for next time."

I get my notebook ready.

"Next time we meet, tell me why agile transformations in companies like WL fail."

"Uh…ok."

"You had a great breakthrough on *environment* today. That's why you have this assignment. Not because you're about to lead an agile transformation."

"Okay." I am hesitant. "Why agile fails."

"Great," Eve grins. "Now let's get on to that bottlenecks conversation."

I close my notebook and tell Eve about how Thad is holding me back with the agile and lean transformation work. I have Rick's blessing and a budget, and I'm ready to get moving. The first agile teams have already gone to training, and are up and running in Ben's area. I've got a core team ready to work. I'm supposed to partner with Thad, but he is never available to talk. He skipped all of our vision work, and then arrived late to our report out on it. Then, he asked us tons of questions that gave us the feeling he didn't believe in the work at all. It didn't feel personal. It felt anti-methodology. All of this makes me suspect Thad is not as ready as I am to go forward, but he doesn't want to admit it to anyone.

Then I tell her about the inaugural C² meeting: Thad is a no-show, and the group found themselves speculating about what he might want of them, what the vision on culture might be, if there even was one.

I share what I learned about bottlenecks, that even though there is an obvious problem with Thad, there could be another side to the bottleneck. I just don't know how this perspective fits into agile and lean thinking.

In the case of the transformation, the other side of the bottleneck is *me*. I'm waiting on Thad. I've done my best to engage him, but he's not involved. So, do I have to wait? Are there things I can do now without him that won't ruffle too many feathers? Are there things that I can do now that will ruffle feathers, but I should do them anyway? Agile and lean leaders are trying to effect change; how much pushing is just enough?

"It's interesting that Thad is a bottleneck in two large efforts," Eve observes. "Most leaders don't understand they are being a bottleneck. Other leaders understand and relish being the bottleneck, because of the power associated with it. Where do you think Thad is on this?"

I tell Eve I think Thad knows he is slowing things down, and he believes he has a really good reason for it. He is probably trying to get as much stakeholdering done around lean as possible, ahead of the work. Although, from what I've seen of his style, he is more likely strong-arming people and causing a trail of destruction. He wants lean to win ahead of agile. Early wins for the new guy, but more importantly, lean wins the attention and the money. Regardless of his reason or reasons, he's not telling anyone what they are, so we are forced to make up our own stories about what is happening. Or not happening.

"Right, Joel."

"There's more."

Eve raises her eyebrows in surprise, but her face tells me she knows there is more.

I tell Eve that I keep thinking about the other side of the bottleneck. So I'm waiting for Thad. Couldn't I just forge ahead?

"Eve, you and I talked about this a while ago: what I think I need permission for, versus what I can just do. Couldn't I tell him, *meet me on this day to review the plans with the team I've selected? If you can't make it, we are going ahead anyway.* I'm happy to update him when he has a chance. Aside from feeling bold, it feels like career suicide," I sigh.

"Wow, Joel, you are in a really great place."

I chuckle, "Yeah, right."

"I mean that. You have learned so much, that you can see things now that you never would have, even a month ago. Or, you would have misunderstood them. I know it's uncomfortable, but you should feel really good about your observations and questions."

"Then there is the thing about C². Now it seems like that committee is a moot point. We are going agile and lean. We will experience a change in culture—"

"If your transformation goes well, you will," Eve cuts in.

"Well, yes, I expect it will go well. I have to plan for the best, right?"

Eve nods.

"So, C² seems like a risk, now."

"How?" Eve asks.

"It's a risk to overall company credibility. Does our organization really stand behind its initiatives? Eve, I can picture the C² work getting off and running, but then will falter, like every other corporate-wide initiative we've had like this. A few years back it was *Winning Together*. It was a big splash for about six months, and then it died. So employees are no better off, and leaders in general will go on thinking their employees don't want to change."

Eve agrees. She says the culture work is premature to the transformation work. It's going to send people off in one direction, and it likely won't be the same path as agile and lean. We don't know this for sure, but in her experience, there is always confusion when this happens. No one wins and no one really knows why. It puts our leaders in a tough spot, because they don't know what work has the greatest priority, and they don't know which behaviors to model. So all of the work becomes important, and they become inconsistent with what they say and do. Teams struggle with the leader's problems, which now manifest in their work.

She also believes there are some small wins that will happen amidst the confusion, because anything can be made better when people are motivated to do it. These wins may stick. The people who experienced the win will feel excited to make more changes, and sometimes they will. Often, though, they run into impediments that hold back any further change. Things seem as hopeless as they were before the initial breakthrough, and they get discouraged. Some will never try to make a change again, reverting to the old mode of waiting to be asked to do something. Some may leave the company in search of that motivated feeling they used to have. Employee surveys and interviews may reveal this, but the company will again point to the employee as the problem.

I blink. I get it, and I'm overwhelmed.

"You only wanted to know if you should do something about your end of the bottleneck, and we went real deep," Eve smiles. "Sorry about that."

"Yeah," I say slowly. "I didn't know why this committee work felt bad, but now I do. I want to stop the work, but I don't think it's the best move for my career. With all that we've talked about, that's my next question. Should I speak up about this, or just let it go?"

"How will you decide if you should speak up?" Eve asks.

I smile and shake my head. "Answer my question with a question. Thank you for that."

Eve smiles, and waits.

"The C² work and the transformation work are both big. Both make a big impact on the organization. But the transformation work and my relationship with Thad are so very close to home, which is where I should begin. The C² work will probably be there running in the background."

"So, when you think about the bottleneck of Thad for your transformation work, what is your next question?"

"I probably won't say this right the first time…" I pause to organize my words. "What am I willing to push forward without Thad?" I stop for a moment. "No, it's more, *Am I willing to push forward without Thad?*"

"Why are you asking this question?"

"I know I can't wait any longer on the work that has to be done. But there is risk in moving forward. More than being worried about getting fired, I am worried that Thad will retaliate against me for going forward without him."

"And?"

"Uh, right." I sigh. "There is risk in waiting for Thad, and risk of moving forward without Thad."

"Joel, what is the risk of waiting for Thad? Won't that build your relationship with him?"

"It might, but if WL is transforming with bottom-up implementation and top-down support, waiting on one person is…" I can't think of the word.

"The anti-pattern of your transformation," Eve says.

"Yes! That's the word. How can we expect to transform this new way if we do it the old way?" I lean back in my chair. "The other part of the risk is that I don't want Thad to think I am afraid of him. I want the cause for agile to be bigger than politics or the span of power of one or two directors."

Eve smiles, "I like that you don't want to fear him."

"Yes. He's a real challenge to get along with, but that should be different than fearing him wrecking my career."

"Yes, Joel. That is a healthy way to think about Thad, at the moment."

"So I know I don't need permission to lead, here. I am willing to move forward on the transformation work without him. It might be the very first of many big challenges with this whole thing," I put my head down again. I am even more exhausted.

"Joel, are you exhausted?" Eve asks. She knows I am; it's her way of being compassionate from a distance.

"Yes. And I know where to begin," I smile.

"I like that it's an *and*, not a *but*," Eve says.

"Right. We just had this big environment discussion. It's apparent I need to adjust the agile transformation approach to include this philosophy. I don't know how it needs to change yet, and I don't think it's for me alone to decide. I think the small team I organized is where to begin. I'm putting too much on my shoulders."

"It's definitely yours to lead, but not yours to own, Joel. The more you work in this world, the more you will learn that agile transformations are owned by an organization. Although no one at WL other than you knows it. Yet. There is still accountability."

"Yes. And this small team will be a great start for us."

"Focus on collaboration instead of consensus." Eve tilts her head to the side. "Maybe use inspiration from the agile manifesto, and think collaboration *over* consensus. That will be a future topic for us."

"This is a lot."

"Yes, it feels like it. But you are ready. So, I have another assignment for you, Joel."

"Of course you do."

"You have your assignment on why agile fails. In addition, there is someone I want you to meet. Someone I want you to interview."

"Interview." I've done stranger assignments in the name of agile coaching before, so, why not?

"Am I doing this today?"

"Sorry I wasn't clear. No, not today." Eve pulls a sticky note out of her messenger bag, and hands it to me. It's got a name of someone I don't know, and their phone number.

"Thanks."

"Chris is a scrum master."

"A man or woman?"

"A man. And, a very good scrum master. Well, he's done a lot of different agile roles over the years. I coached him long ago."

"What's the intent of the interview?" I pull out my notebook.

Eve smiles, "Very good, Joel."

"I'm getting there."

I'm to interview Chris on why agile fails, and to understand the current state of agile and any other methodology adoption in his company. The intent of the interview is to further round out my perspective on what's needed for a successful transformation by learning what Chris has seen work and what he has seen fail. By learning from Chris, maybe I can avoid making a few mistakes. Maybe I will be inspired to try some new experiments.

"He knows you're going to call him, so you can call any time. He's excited to talk with you."

"All righty then."

Reflections

Corporate environment must change for a transformation to work.

Culture change is an outcome of a transformation.

I am not afraid of Thad, or of getting fired.

I am ready to tackle my first impediment to our transformation.

Tony is still a pain in the ass.

Have I changed since high school? I think I have changed a ton. Mostly positive changes.

I feel like I am changing almost daily right now. Hopefully for the better.

The Interview

Chris and I agreed to meet at J&L's Café, just two days later. I was surprised we were able to meet so quickly; I assumed he would be very busy with his team. I don't know that much about scrum masters, but I know they are a servant leader, focused on their team, and helping them get their work to done.

We have coffee, and a table by the window. I begin by thanking Chris for making the time to meet with me, and then we get down to it.

"You're here to interview me about agile, right? So you can learn what *not* to do?"

"Right," I say. "I need to ask you why you believe agile fails. And I have a few other questions, too."

I ask Chris where he is a scrum master, and if he can't tell me the company name due to any competition reasons, just give me a general idea. He works for a large bank, with many employees and locations around the world. The company is doing very well, but is always on the edge, due to competition and technology changes. Sounds familiar.

"I'm no longer a scrum master in the software development side of my company, but I give this advice to you, and others like you, so I can keep it fresh for myself."

Chris was a scrum master for nine years, at his current employer, and at two other companies. His focus has always been on practical, effective scrum. He gives me the titles of a few books to read by agile expert Mike Cohn, and suggests I read them cover to cover. No skimming, or I will miss out. These books were the foundation on which Chris built his scrum master career.

I thank Chris for the tip, and tell him there is so much information available to us in blogs, books, and videos, that it's difficult to find the truth about agile. And, the truth about why agile fails. Then there are the competing consulting firms looking to engage our company. And there are methodologies out there; some seem over-engineered, while others seem too lose to work in a company like WL. What's right? What's wrong?

Chris agreed, noting he's seen a lot in nine years, including the over use and generalizing of the word agile. It no longer means anything, so now it has to be further quantified, like, a*gile mindset, agile project management, agile-lean thinking.* When people say, *Yeah, we don't do that, we do this, because we're agile,* it causes developers to roll their eyes and disengage.

"So, why do you think agile fails, Chris?"

"Having product owners from the actual business unit is key to successful agile. It's rare for a technology product owner, or anyone not from the business, to be successful in the role. It takes a special, committed person to pull it off."

"Because they just aren't close enough to it?"

"Right. So they are determining priorities for software applications for their teams and the business."

"I can see that happening, Chris. Instead of clarity, this situation usually creates disorganization and confusion."

"Yep. Then, if there is a person from the business in the product owner role, they may not have been trained on how to be one. So they are set up to fail as well. Agile doesn't work for them. End of story."

I take notes, and already I'm thrilled to have met Chris and listen to his wisdom. Once again, Eve pushes me into something at just the right time.

Chris believes another reason agile fails is that leaders are led to believe agile saves money, and gets teams to go faster. It's a sad myth that agile project management processes make the work go faster. The use of scrum and proper agile estimating and planning may show that the project is much bigger, or more complex than originally intended. Something that the waterfall methodology fails to reveal quickly every time.

Chris asks me if I've ever lead or been a part of an 18- or 24-month project that seemed to be going well, and then only at the bitter end, did everyone realize the project didn't meet the mark? The kind where in the end, the customer didn't like any of it, or it was outdated. Yes, of course, I have painfully been involved in that type of project.

"This is the deal: the end-to-end development of a software tool may have the exact same timeline and budget if it is done using waterfall compared to scrum. The difference is that the iterative and incremental development method helps to get the *exact* software that the clients and customers want and need at that moment in time. Through iteration, smaller chunks of value can be delivered sooner, over the life of the project or effort, not just at the end."

"Huh."

"Also, the earned value of scrum versus waterfall is through the roof. I know earned value has some controversy on how it is defined, so hang in there. For what we're talking about here, I mean it in a conceptual way; earned value is something the customer will pay for. Anyway, scrum just blows the doors off waterfall in that regard. And, it's not just that you are planning and incrementing features, it's the practices. You have to do the planning, embrace the use of velocity and get the backlog sizing right, not wrong."

"I see. At least, I think I do."

"Rushing through it is another reason agile fails. There is no perfect way to begin any sort of transformation, but there are many ways to get off to a bad start. One of the worst ways I've seen it happen is being in such a big hurry, that you skip over some of the most important things."

"Like vision?"

Chris nods yes. His personal observations and the feedback he's heard from others is that there is a great deal of disorganization, confusion, lack of clear goals, inept or phoned-in product management, lack of understanding of requirements, lack of business knowledge, and lack of project process. And yet, everyone has the pedal to the metal to push through.

"Phoned-in product management?"

"The product manager is remote from the development team, or the other way around. All of the Skype connections in the world can't save the fact that if the team creating the product isn't in the room once in a while with the product owner, there will be gaps in understanding."

"Oh. I can see that happening," I say. "I would think there is a way to compromise?"

"Of course. That is, if you understand that there is a need to be together."

"Oh."

"Right. These phone-in situations, most don't see anything wrong with it. The team doesn't like it, the product owner doesn't like it, but no one thinks there is anything that can be done about it. Maybe that's the case, and maybe that's one of the teams that can be agile, but maybe not scrum."

This is a fascinating topic, but I need to remember my assignment. Maybe I can meet up with Chris again. I want to circle back on the urgency part of this conversation. I feel that's a strong part of what's about to happen at WL.

"There is a huge temptation to get up and running at my company," I admit.

"Exactly. Once the decision to go agile is made, there is a flurry of activity. This is only bad when there is no vision and strategy organized around that vision. The other thing is, when you begin, it's all going to feel too fast. You're not going to feel ready, mostly because there are so many unanswered questions."

"In the fog."

"Exactly. But, if you have a vision and a strategy, you *will* have a way. This is what people can cling to when everything seems to be in a fog. And, when things get really difficult."

"So, does this mean that leaders are a reason that agile fails? You can have a vision and strategy, but if someone or *several someones* don't lead with it, don't buy into it…"

"Right. I've read about it, heard about it from friends, and through networking at conferences. So many leaders believe agile isn't for them, period. It's for teams. They grabbed onto just one part about bottom up implementation and top-down support."

"Huh. That's what we're doing."

"Of course you are. There is nothing wrong with the approach, unless leaders don't fulfill their end of the deal. You've heard of the breakfast analogy, right?"

"No."

"Just as well. Basically, think of a breakfast with eggs and bacon. The chickens *contribute* to the breakfast, just like leaders, but the pigs are fully committed, which is our teams. It pretty much says that you're trying to get them to commit to a suicide mission and that their sacrifice will not only be unrewarded, but only appreciated for a short time. The analogy is de-motivating to intelligent or imaginative people, so I don't recommend using it."

Chris is right. Leaders definitely have less in the game. Although I'm feeling all-in as a leader of our transformation, I'm not the one being collocated with perhaps new team members, on a new product, with deadlines to balance with customer feedback. I'm not the one standing in the front of the room every two weeks showing what I got done, and asking for feedback. Wow.

"In most organizations *accountable* just means that you have to explain yourself to someone if things go wrong and that you might get yelled at, but there are no real consequences. The actual consequences when things go wrong will be suffered by the teams. But that's okay, because they are resources to be consumed, anyway. The root cause of the problem is the incentive systems associated with the hierarchical structure that puts the value stream on the bottom."

"This is great, Chris. Disturbing, but great."

"Speaking of training, agile fails when there is no follow up coaching, and there are no checks and balances. It's not about the teams being smart or dumb. Many companies go to training, but do not hire coaches after the training. This is a problem for two reasons. First, if everyone did get trained, that's great. But then they return to work, and begin practicing agile and scrum. Then guess what happens?"

I think of Ben's teams. It can't be easy to be the only three scrum teams, when everyone else is waterfall. "Uh, they run into friction with the rest of the organization? The environment?"

"Right. Fresh out of training, with no protection from leaders or agile coaches, the team is likely to quickly slip back into old ways of working. The environment puts tremendous pressure on these new teams, and newly trained leaders."

Chris keeps going. Scrum has evolved in practice over the years, so that there an essential tension and conflict between the scrum master and the product owner. It's really for the good of the product, and the good of the team. The scrum master not only handles the project management tasks, but also represents the team. The scrum master is the one to say, *You will not take advantage of my developers and force them to work 80 hours a week,* and then shows the product owner and stakeholders exactly when things are expected to get done, by volume, based on the size of the backlog and velocity, so there is no confusion.

When there aren't many scrum masters and product owners trained and supported to operate this way, the natural predations of the product owner can win out, causing the team and the product to go off the rails. And eventually, agile will be blamed for it.

"Wow."

"The scrum master also facilitates collaboration between the product owner, the person who wants something built, and the team, the people who will build it for her, so that everyone has a common vision of how they are going to solve a customer's problem. This is never an easy task," Chris says. "I've heard people at my current company say, *Since we are experienced with scrum, we don't need scrum masters.* This statement actually shows that the exact opposite is true. They have a fundamental misunderstanding of the importance of the roles in the actual execution of the process."

Our time is almost up. It flew by with this incredible conversation. I could talk with this guy all day, but we both have to go. I want to make sure I correctly captured Chris' key points before we part. I look at my notes, and then read aloud the following list to Chris:

Why Agile Fails
Product owners who are not from the business.
Belief in the myth that agile saves time and money.
Launching agile and scrum teams in haste.
Agile roles not embraced or skipped over.
Leaders misunderstanding their role and commitment to agile.
No vision and strategy for people to connect and hold onto when times are tough.
Not training everyone.

"That's a great summary of what we covered, Joel. It's not complete, but it captures the big picture. If I think of more, is it okay for me to drop you an email or a phone call?"

"Yes, please. I would like that very much. I really want to give this launch my best. Eve was spot on connecting us so I could learn from you. I have two more questions: what do you do now? And why did you stop being a scrum master?"

Chris found a new job as a product manager in another part of the company. He said he enjoyed being a scrum master, and there was decent money to be made doing it, but he had an opportunity that was the right work at the right time. Now that he's working on *the other side*, he has some empathy for leaders and product owners. He's hoping that his experience from both sides will help him make a career change he's been thinking about for a long time: to break out on his own, and become an agile coach. His wife has a stable job with good benefits, so he could work as an agile coach under contract for different companies.

Huh.

I share with Chris that we are looking for strong agile coaches, and that perhaps if he's ready to make the leap now, there might be an opportunity for him at WL to help from the ground up. Chris' face lit up. I had to confirm with him that I was serious, that I heard enough from our time together, that I would like my transformation team to meet him. I couldn't promise anything, because I'm not in charge of who we hire. So please don't quit your job, just meet with us. See if it's close to what you are looking for, and we'll see if you are close to what we are looking for. Listen to me, recruiting my first agile coach while on assignment from *my* coach. So many days, I feel like I really don't know what I'm doing. Right now, this feels like a good thing. A sure thing, in my foggy, uncertain world.

Did Eve plan this?

We shake hands, and I head back to my office with fresh perspective. It's the best 90 minutes I've spent since my coaching session with Eve. I loathed the idea of adding just one more thing to learn or discover to my plate, but this was a win. Once again, Eve pushes me just enough, at just the right time. Yet it was up to me to say yes to the assignment, and go for it.

I can't wait to share this list with the transformation team. The part about why agile fails, mostly, and a little about considering Chris. I have to be very careful with that part, so they don't think I'm forcing someone on them.

Then there is my (now) long list of why agile fails. My notebook must weigh 300 pounds! There are so many ways agile can go wrong... In a large organization, is there any way to not screw up agile? Maybe that's it: an agile transformation will never be perfect, but it can be healthy. Imperfectly healthy. If we begin the right way, amidst the imperfection, we have a fighting chance. Have a vision, lead with the vision and the strategy, pace ourselves, and get the right people involved from the start. Then, incrementally sense and adapt to change for the better.

Simple enough, I laugh to myself. Who would ever take on the challenge to lead something like this?

Reflections

Large organizations struggle to do agile well.

Leaders are not as committed to working agile as teams are, which alters their decisions.

Teams have a huge commitment to change when on a scrum team.

Agile desperately needs business partners.

Business partners must be fully committed to their product for agile to work for them.

Agile takes the blame for other preventable dysfunction.

Chris' List on Why Agile Fails

Product owners who are not from the business.

Belief in the myth that agile saves time and money.

Launching agile and scrum teams in haste.

Agile roles not embraced or skipped over.

Leaders misunderstanding their role and that a commitment to *be agile* means everyone.

No vision and strategy for people to connect and hold onto when times are tough.

Not training everyone.

Leader, not Owner

I'm out for a run before work the next day. I'm hoping the time alone will help me clear the swirl in my head from yesterday's session with Chris, and my coaching session before that. Then there is my anxiety about Ironman Boulder. I can't believe I'm going to do an Ironman in four weeks. Four weeks! I think it's healthy anxiety, but it's uncomfortable just the same. So there is a lot of swirl going on that needs to stop. I have a training plan for my race, and I have my reflection notes for my work with Eve. Now I need time away to be present in right now. I need silence, and I need clarity.

I try to focus only on the things I can sense in this moment. It's very difficult. There is so much to do. A few months ago, I didn't know if I wanted to stay at WL, and now I'm leading an agile transformation. I have a small team to help me, but they could only commit a fraction of their time to the team. Will that have to change if we are serious about agile? Does this work need more money or more people or both?

I have my phone on an arm band in case of an emergency, but I'm not listening to music, not answering any texts, emails, or social notifications. This time is mine. I notice the warmth of the sun, the sweet smell of the air, and the sound of my feet on the path. And birds. My thoughts bounce all over the place for a few miles, but then I'm able to find enough calm to focus on my breathing. I'm folding into the moment, feeling lighter as I let go. I'm deep in the nature around me, that I barely notice the hills on the trail.

The magic is over. Seven miles ticked by in a flash, and I'm back at home. It was great while it lasted. I had some great moments of clarity. Driving to work, I can't ignore how Ironman Boulder and leading the agile transformation at WL are converging in front of me. They are not exactly the same; if I fail at Ironman Boulder, I won't lose my job. I will be disappointed, but my livelihood is probably very safe. And, Ironman is just one, long day. When I finish, that's it. Although I hear from a lot of other

Ironman finishers that once you complete one race, you are hooked, and want to compete again. The finish is the beginning. I'll pack that away for now. I can't think beyond this huge day waiting for me.

Leading this agile transformation has a little more risk. If it fails, there will likely be side effects to my career. Eve's words pop into my thoughts that *the transformation is mine to lead but not to own.* And, that likely no one else at WL is aware of this. And it's a long-term commitment. I've learned from Eve, and from my scrum training, that Agile is a journey that is different for each leader, team and organization.

Both have an annoying player in my way. Well, I suppose it is about perspective. Tony is annoying, but I can choose not to have him in my way. I don't have to accept his offer to ride with him, no matter how relentless he may be about it. And if I see him on race day, I should be busy following my coach's plan to execute my race, and not worry about anyone else. It's very tempting in races to get caught up in what someone else is doing, or start comparing myself to others around me. If I do this, I will lose my focus and not be able to execute my plan. My coach believes my focus is so important to my success on Ironman day, that I had to come up with a mantra: *my race.*

After talking with Eve, I'm positive I'm putting Thad in my way as much as he was put in my way. We are both leading part of the transformation of WL. Thad has the lean side of things. Eve didn't like how the methodologies were divided like work. Dividing agile and lean between two leaders who have friction to begin with is not good. Eve called it purposely taking the long way on a trip.

She said the irony of it all is that Rick is creating exactly the same kinds of silos that I am (and Thad is) charged with eliminating. And, the reason these two silos were created is because this is how WL leaders (and many US companies) build their delivery systems. The system leaders create is the root cause of everything. Huh.

Eve is ahead of me on that, so I can't worry about it now. Today, my biggest problem is wasting energy second guessing my decisions that shouldn't involve Thad, and waiting on him for things that I probably don't have to. That is the comfortable way to lead. Make someone else responsible. I knew better before I met Eve, but our discussions now give me specific guidance on why this behavior needs to stop. If only Thad were getting the same coaching as me…

I don't feel like I know what I'm doing in either place, but I'm going to persevere. I've put myself inside two giant goals that I've never done before. I'm going to trust the coaching I've had in both, stick with my gut, and learn as I go. That's the convergence.

I'm in Thad's office for a weekly alignment meeting. We haven't had one in two weeks because Thad's been a no-show. Today, he was actually in his office and ready for our meeting. I'm immediately suspicious, but glad he is there. I've mentally prepared myself to face him with the bottleneck discussion. I don't want to delay it another day.

"Come on in, Joel." Thad is looking at his laptop as he greets me.

"Let's talk transformation teams. Now that we have a transformation vision for WL, we are beginning to build the team," I say.

"So glad you brought that up…" Thad tells me that he has organized a Project Learn team, complete with people and project managers. They are starting with their first RIE in about six weeks.

"What's an RIE?" I ignore the fact that Thad still used Project Learn.

"Rapid improvement event, Joel," Thad says with supreme confidence. "We can talk about it later, if you want. Basically, it's a big process improvement event, designed to reveal waste and customer value."

Thad also shares with me that, by the way, he and his team are using RIE to refer to the work, because they want to *shy away* from using the word lean. I

90

can't believe this guy. And where is Rick in all of this, waiting for us to collide, so we can have it out? Or, is he oblivious to it because we've been hit in the gut by our competition, and by regulatory fines?

I don't want to argue with Thad about using RIE vs. lean, and lean leadership. But if I ignore this comment today, I'm setting a precedent. So I tell him, and offer that this is a great example of how we could work together better. I have nothing to lose.

Thad looks to the ceiling, with an index finger to his cheek, "What does that have to do with agile?"

"It has to do with leading transformation," I say. "Let's try to work together, Thad. Before either of us really ramps up our work."

"We've got some traction in the operations department," Thad chuckles. "The recent regulatory fines are actually working in our favor."

Thad and his team are advancing the transformation without anyone knowing it. It's wrong on so many levels, but I have to step past all that right now, because of the glaring problem of not being aligned.

We are going to keep talking about him and his RIEs if I don't get this conversation back on track. He's banking on me having a stronger curiosity about his work than my intent. Not happening.

"Thad, it's great to hear about your successes. I don't want to discount them, but I do want you and I to be aligned, and lead teams that can work together for the overall benefit of WL."

Thad pauses his typing and looks at me for a brief moment with his icy blue eyes. "Isn't that what we are doing right now, aligning?"

"It's informing." I say, resisting the urge to tell him what he just did. I don't need to say it because he knows exactly what he's doing. "Aligning would be sharing and discussing ahead of the actions. Maybe even some planning together. And, it would be grounded in our common vision."

Thad shrugs and returns to typing. "We are aligning in this alignment meeting. And, we have shared a common vision. I think we are doing great."

This is going nowhere fast, but I have to say it. I have to challenge him to support our team vision. Think about intent... "Aside from the minor detail about using the word lean, the transformation team has a new overarching vision and name."

"Really?" Thad looks up at me, shocked.

"Yes, remember the one that was built from feedback collected from our practitioners and leaders: *We are continuously learning how to better create and deliver ways for our customers to connect to the world. We are...Becoming.*"

"You don't have a vision; you have a *name*, Joel." Thad checks his phone. Does he have any idea how he uses his devices to mask his social awkwardness? Whatever Rick sees in Thad, it's not because he can hold a decent conversation about work. Now it's looking like he can't collaborate. A lone operator driving for results. A walking time bomb of sorts for the rest of us. And he does a tremendous job of hiding it.

I am silent, trying to keep my jaw unclenched and relax. I act as if I'm waiting for him to finish what he's saying. Focus on *intent*.

Thad looks up, "You know, we have some big plans for lean at WL. We can't wait around for a transformation team to be built. It's roadblocks like this that hold us back from transforming, Joel."

Now I am the roadblock. Of course I am.

"I have a lot of experience in this space," Thad sighs. "That's why Rick has me here at WL, and why I am leading the charge for us. I am able to discern between an honest roadblock and garden variety friction."

So I am a roadblock, not garden variety friction. Yeah, I can try all I want to keep my focus and keep the intent of the meeting, but it ain't going to work.

"We are leading it together, Thad. We each have a responsibility for this transformation."

"You can think that if you like," Thad chuckles. "But everyone knows lean is the only change that lasts, with or without a transformation team."

Ugh, it's impossible to have a sensible discussion with this guy. I am stepping around so many things to avoid further manipulation, that I'm losing my footing. And, it's exhausting. One more time, I can try to push my intent.

"We should take time to get our teams together, and at least review the feedback from the WL teams and leaders surveyed. It's a very customer-centric approach that will help us sustain our change in the long run. I think we can both agree on that."

"What are the details? How does that look, Joel?" Thad asks, for sport.

"We should talk about how we can be better aligned than we are right now," I say slowly. Stick to the intent, one, last time.

"Look, I've got a lot going on Joel. This is big baseball," Thad motions to his laptop.

Whatever that means.

"All the more reason we should work to be aligned."

Thad ignores me, and shares that tomorrow, he and a few other operations directors are traveling to a company in our industry to research their transformation. Thad tells me more about the company and their lean transformation, but I'm barely listening. All this time, I was waiting for him, and he's been moving ahead like a cowboy. Rick must know about this. Then again, why would it matter that Rick knows? He hired Thad to drive forward the transformation at WL, and drive he is. Without any partners.

Thad continues, stating it's going to be a great opportunity for us to learn from a company that has gone ahead of us in their transformation journey.

Their products aren't as strong as ours, but they will be a good sounding board for us. We can really leverage what they've learned, so we don't make the same mistakes. We have other trips lined up, but this is the one Rick wanted us to visit first.

Well, there. Rick is aware. The longer I work for Rick, the less I understand him. It's also surprising that I didn't hear about this work from someone else in WL. Even though Thad's traveling with operations directors, our paths cross on occasion, I guess. It really doesn't matter now. Thad is the problem, not others who are involved.

At the end of his glowing description of what he and his team plan to do on their trip, Thad assures me, "I knew you were busy with your vision, so I didn't invite you."

Alright. I've been focused on intent, but nothing is happening. I need to speak up and try to right this relationship. It's now or never.

"Thad, you're acting outside the partnership we established to lead the transformation of WL. We had a plan, and now you're moving in a different direction. When you act outside our partnership, it causes chaos for me and for our teams, and impedes WL's transformation. I want us to get back in sync, so we can create a sustainable change."

Thad doesn't look up from his laptop. "Joel, you're overreacting. None of the teams are in chaos," Thad smirks. "Well, none of *my* teams are in chaos."

Thad waits for me to take the bait and grab onto his chaos accusation, but I don't. I look at him and wait. This is uncomfortable, but because Thad rarely makes eye contact, it's much easier to be silent than usual. My waiting pays off, because Thad picks up where he left off.

"We are a partnership, and we are in alignment. I hope we can continue to leverage each other, when strategy dictates it. Your agile work doesn't need the same focus, so it's plausible that we can bifurcate on occasion."

I wait to say anything to him, because all that comes to mind is *bifurcate my ass*.

"It's really going to be great, you know," Thad proclaims. "You just have to believe in yourself, Joel. It's a difficult journey, but you will be fine. Leverage me any time for guidance and advice," Thad sighs. "This is my wheelhouse, Joel."

Thad looks at his watch. "Oh. I have to go meet with the ops team. We are interviewing some lean coaches to help us get the transformation going."

I stand up and open the door. "Well, I'm on my way. Talk to you soon."

"Thanks, Joel. It will all be okay."

I leave with a bad feeling deep in my stomach, for so many reasons.

"He's hiring more coaches? I thought those two at the RIE were there just for the event." Meryll gasps. "What is going on around here?"

"He's driving the transformation at WL, that's what," Jack sighs. "Action wins!"

"Fire, ready, aim!" Meryll sighs.

I nod in agreement. "I feel like a fool. Here I was thinking that he wasn't engaging mc because he was overwhelmed. You know, with him being the new guy, he was spending a lot of time just getting his bearings. Man, was I wrong."

"Are you going to approach Rick?" Meryll asks.

"I don't think I have a choice," I say. "But first, I need to reflect on all of this. There are so many moving parts."

Jack nods, "Yeah there is a lot going on here. Including your original goal of moving forward on stuff that you don't have to wait for Thad."

"Yes, and you don't want to look like you're overreacting or in a panic," Meryll says.

I sigh. "So, what do you think of my next step? I think it should be to meet with my team as intended, and get rolling on things. I'll keep inviting Thad, even though it seems like he won't be involved."

Meryll shakes her head, "Yeah, until he is. We all know this cat well enough to predict that he will drop in to your work and your team's work at the most inconvenient time. The better you are at being open to it, like it's expected, the better for everyone."

"Meryll's right. You're gonna have to keep trying to align with him, because if you don't, someday he's going to play his hand, and make it look like *you're* the problem. You're gonna get nailed for not being aligned."

"Of all the things you two are telling me, and it's all very good, I have this gut feeling that I'm already going to get nailed."

"Nothing you can do about it," Jack shrugs.

"Come on, Joel." Meryll huffs. "You haven't done anything wrong. You're just wound up."

"Yes, I am. But there was something about the way Thad reassured me that everything *will all be okay.* Something snapped in that moment, and gave Thad the upper hand. It was creepy," I sigh. "Now I'm just mad at myself."

"No way," Jack is firm. "You gave him feedback and used good words. Most people would have either walked away too pissed off to say anything, or they would have bit Thad's head off. You gave him intentional feedback."

"You had to do it, Joel." Meryll stands up, and starts pacing.

"Not bad, Meryll," Jack acts impressed. "You didn't begin pacing the floor for 15 whole minutes."

Meryll stops pacing, rolls her eyes, and then paces again. "This problem has been simmering for a while, Joel; you have been hounding Thad to attend your meetings, read your customer feedback, and support the vision and strategy. He's ignored all of your overtures. You even checked with your coach on the best route."

"Thanks, Meryll. I've got the best laid plans for this transformation," I sigh.

"So, what have you learned so far?" Jack asks. "That's what my coach Steve would ask me."

I smile, "They are having the desired effect on us, aren't they?"

"Don't I know it," Meryll mumbles.

I've learned that the bottleneck problem was important to me and my team. Now that I've released myself from waiting on Thad, we are free to get moving on our plans. Also related to this is how I told myself a story about Thad, without any facts. I assumed he was busy or working on getting to know WL culture and leaders, when in reality, he was forging ahead with secret transformation plans.

Thad sees the transformation as only two parts: lean and agile. He believes he has one half, and I have the other. Well, there I go making another story without the facts. Anyway, we are supposed to be one team, one transformation, to become the new WL. The one that focuses on the most important work at any given time, that doesn't compromise our customer's data or their experience, the one that connects them to the world…

I also learned that Thad is an action guy, just like Rick wanted him to be. Rick wants action more than he wants collaboration, even if he won't admit that.

I learned to give feedback when the risks are high. I'm very happy I gave Thad feedback about the situation. I felt so much better after I did it, for two reasons: first, this problem needs to get out in the open and discussed. Check. Second, it was a big challenge for me to speak up like that. I'm comfortable and experienced with advocating for myself, but this situation has lots of risk because of the work we're doing, and Thad's personality. A few years ago, I would have absolutely ignored this problem.

There are two things still bubbling up to the top, so I can't say I learned anything about them yet. First, Thad is probably going to use my feedback to somehow politically hurt me. My brain anticipates several different ways he'll do it, but I'm going to work on letting go of that. It's out of my control.

The other issue still forming is with Rick. What does he know about Thad's cowboy ways, and what does he care about them as they relate to the overall success of WL's transformation? It's not even about lean or agile; it's about his leader team acting like leaders, or like leaders of super-optimized silos. Maybe those aren't the right things I'm going to learn. Maybe it's going to be about how WL and leaders like Rick are addicted to action.

Speaking of lean and agile, that's another thing that I'm probably going to learn about over time. It didn't really cross my mind that lean and agile could work against each other until Eve made that side comment about giving two different methodologies to two different leaders.

But it's deeper than that. Until I met Eve, I was mostly resistant to lean. I thought it was just more quality BS that was forced on us, and wouldn't actually help anyone. Then Eve's coaching got me to see the value of lean thinking, especially lean leadership. Add that to what I've learned from Eve about agile, and I don't see much friction between the two methodologies. They can work together very well.

So there is another thing I learned about lean. The only friction with it has been created by people. At WL, we've been told not to use the word lean. The corporate communications people and senior leadership determined that lean is associated with layoffs, and should be avoided. They don't want people working in an environment of fear. Yet, by not being open and forthright, they are damaging trust in the organization, which will create fear. I'm guessing enough people know about the forbidden lean word, it's likely a problem already.

Thad's arrival at WL created friction with lean and WL. He was brought on as a lean expert, albeit from a manufacturing background. When we have seen him around, he's been using lean as a hammer, rather than as a methodology or mindset. Forcing his thoughts on the room, and speaking in absolute terms are just a few of the ways he's hammered away on us. I'm no expert in transformations, but I know enough from leadership training that it's bad to force yourself and your initiative on people.

"That's a lot of learning," Jack raises his eyebrows. "Too bad it's all mostly as a result of crappy behavior."

"I will be all right."

"Of course you will, Joel," Meryll stops pacing. "We know you don't want sympathy."

Jack chuckles, "Yeah, we just want Thad to go far away."

"What are you going to say to Rick when you see him? I mean, how do you even begin with this?" Meryll asks.

"Great questions, Meryll. I don't know, and I don't have to know today. Maybe all I need to do today is to plan the next meeting with my team. Which will be today. I'm just going to have it, and see who can attend."

"What about the group traveling to that other company?" Meryll asks. "Don't you want to contact them to learn more about their visit?"

I tell Meryll I'm not going to contact them. Of course I want to, but there are bigger problems than this. She knows it; she's just reacting to the story.

"I can tell you a few things I've learned from this," Meryll sits down. "Thad's turning into a lone wolf. And we can't trust him as far as we can throw him."

Reflections

Leaders do not own it all.

Being intentional and courageous helped me navigate a very manipulative conversation.

I believe Thad wants to sabotage me and my work.

A methodology war is in progress.

Repeating Problems Become Culture

I'm in the kitchen, having a late lunch after my last long bike ride before Ironman Boulder. I used to think any ride over 15 miles was long, but now my baseline long ride is 75 miles. That's what I completed today, followed by a 30-minute run. I was able to avoid the awkward phone call with Tony this week, and just ride with some friends I met through my coach. It was a stroke of luck: Tony contacted me and informed me that he had to fly to Houston on business, and wouldn't be able to ride this weekend. He hoped maybe we could get together next weekend.

If I'm asked next weekend, it might not be so bad to ride with Tony. With the race just 14 days away, the workout plans will be short. Any time with Tony is too long, but I'd rather put up with a 50-mile ride with him than a 100-mile ride.

I told my coach about Tony, and he laughed. He said it's about time I have the challenge of a competitor getting under my skin. Here I thought I was challenged enough training to just finish my first Ironman. No, my coach said the mental muscle needed to succeed at an Ironman triathlon is built from all sides, including annoying competition. So great, I can check the box on that: riding with The Mouth will help me become a well-rounded athlete. Just so Tony doesn't know that he's helping me become *anything*.

I chuckle to myself about Tony. He's not the worst person in the world; I just can't stand to be around him. My coach said that the sport of triathlon has a lot of good people, and also people like The Mouth. He said these types of athletes can get in your head and cause problems with your self-esteem during training, and really mess with you on race day. In fact, my coach has seen many athletes have meltdowns in races. Many of them were due to a rival or bothersome acquaintance passing them during the race. Getting passed in an Ironman triathlon isn't always permanent; it's a 140.6-mile race. But if you lose your mental focus at mile 50 of the bike, you could be finished before you're even halfway through the race.

The fact that these people are annoying isn't the problem; it's what we *do* with the annoying. My coach said the athletes who are able to pass off or pack away the *annoying* have the most success and satisfaction on race day. Those who let the *annoying* permeate their thoughts and their race have poor performances. Which can also mean they had a good athletic performance, but their special day is spoiled with the memory of how the *annoying* got to them.

Maybe the strategy to find a place for the *annoying* can be applied to more than just Ironmans; Thad is far beyond annoying, and I have to work with him for hours a day, but maybe this strategy can still work.

Yesterday's dinner leftovers taste like a dream: pulled pork on a tortilla with some slaw. Just then, my 17 year-old daughter, Caroline, walks in, returning from work. She is a lifeguard at Splash Water Playground, or just Splash. It's a water park connected to a hotel. Both hotel guests and locals can buy passes for the park. Our family has been there many times in the past for birthday parties.

It's the kind of waterpark party where the kids have cake, run around the waterpark and swim, and then spend some time playing a few games in the arcade. They finish agonizing over spending their prize tickets on worthless trinkets, like spider rings, and parachute army men. Caroline grew up going there, and so we weren't surprised when she wanted to work there. It's just three miles from our house. In the world of running people around all over Denver, a first job three miles away is perfect.

I ask Caroline about her day. There is never a dull moment, so I was not surprised when she huffed about being starved, and having a killer headache.

"Did you forget your lunch? Wait, aren't there snacks in the break room for you?"

"Well yeah, they serve us lunch. It's made by the hotel chef, just for the water park employees." Caroline puts some leftover pulled pork in the microwave. I smile to myself. She barely touched the pork for dinner last night, opting for a mostly vegetarian taco of slaw, cheese and guacamole. Today, things are different.

"The problem is, if they are behind on giving breaks, it doesn't matter what's in the break room, because I can't eat." Caroline sits down next to me, and digs into her food. I love our talks together, even when they are to discuss problems.

"You worked the morning shift; how could *they* get behind on breaks? And, who is *they?*"

"*They* are the people on the team working with me. Other lifeguards."

Caroline then tells me that three people did not show up for work this morning. Apparently, there was a party last night, and these three guards were out late drinking. The leaders scrambled to get someone else to come in, but with a teenage staff, many were not answering their phones on a Saturday morning.

"Were there enough guards to keep the park open? To meet your safety standards?" I ask.

"Not at first." Caroline takes a large bite of her taco. She nods her head as she chews, and looks incredibly happy eating the recently-shunned pulled pork. I know better than to point out how her eating standards have changed, so I just enjoy it from a distance.

"For the first hour, we were not up to safety protocol. But the leaders said we would still open. They told us that the general manager of the hotel wouldn't delay opening because the hotel was sold out the night before. Dad, that doesn't make sense to me. If you know you're going to have a bunch of guests swimming, and you don't have enough lifeguards to meet the safety standards, you should make an adjustment."

I nod, "The GM was willing to bet that there wouldn't be a ton of guests right away, and also willing to bet that someone else from the team would come in to work."

"Right." Caroline scrapes her plate clean.

"So, was is really busy that first hour?"

Caroline tells me that the first 45 minutes of being open were calm, but then it ramped up fast. The leaders were only able to get one other person to come in. So without enough lifeguards, there could be no rotation for breaks.

I ask about the lead lifeguard. Can't that person give the breaks? Caroline said no, the leader was too busy making phone calls to get more people to come in, and then didn't have the right gear to do the job. I asked why the lead guards don't just have gear at the water park, and Caroline said they just don't do it that way.

Caroline also told me about Rhoda, an older woman (Caroline's description was less flattering) who has worked at the park for seven years. Caroline said that Rhoda is the only employee who doesn't have to work any Saturdays. The rest of the crew must work at least three Saturday shifts a month. So with the forced Saturdays requirement, people agree to it, and then just call in sick when they don't feel like going. But not Rhoda. She can work whenever she wants.

I want to talk more about Rhoda sometime, but I want to get back to the breaks problem.

"Caroline, I'm sorry you were so stressed out today."

Caroline shrugs and finishes her water, "It's work. It's just the way it is. I'm just glad I only had a six hour shift. I hear this happens sometimes when people have an eight hour shift. That just doesn't seem right, Dad. It's just a water park. We're not saving people's lives in the ER, or something."

"Good observation," I nod. "So this has happened before. I can see the GM's point of view of wanting happy customers, and as few problems as possible. But there are risks of doing it this way."

"There sure are!" Caroline opens a yogurt. "I don't think the leaders or the GM care about us at all. They are only concerned with opening on time, and having everything up and running."

"Was there any acknowledgement that you were missing a break? You know, a *sorry about that, thanks for understanding* type of thing?"

"No. At least, not to me," Caroline shrugs.

"Huh."

"You know, Dad, what bothers me the most is that this happens often."

"It's part of the culture there."

"Yes, it's just the way it is," Caroline sighs. "And that the lead lifeguards don't step in to help. If I had that job, I would want to help, and to make things better. I would not be able to just hide in my office, while the team is barely able to get a bathroom break. No wonder people don't show up."

"What else would you do?" Now I sound like Eve.

"I would talk with the GM about the risks of employee safety and morale. Since the GM won't agree on keeping the park closed until we have enough people, I would propose an idea to get the lead guards involved. Once those lead guards are finished trying to call people to come in, they can help give people breaks. And you know, maybe if they treated the team better, fewer people would be no-shows."

Caroline shared more ideas with me on how she would run the park differently. It was refreshing to hear her thoughts on something other than friend drama and homework. I praised Caroline for thinking of others, for considering how others might feel about the situation. We talk about how difficult it can be for leaders to make work a positive place if the leaders above them, like the GM of the hotel, don't value it. Oh, the irony here.

I ask Caroline if there is any consequence for not showing up for work on a Saturday morning. She said there is, but that's loosely enforced. It depends upon how many new people there are, because the experienced people may not show up all the time, but they need them when they do.

"I knew this job was going to be crazy, and that it wasn't the highest paying job around. I picked this one because it sounded like fun." Caroline cleans up her eating space on the table. "Today was not fun."

"Good things to think about, Caroline. First, how you would do it differently, and next, what options you have given the current state."

Caroline pauses for a moment and then smiles.

"What?"

"I can tell you're being coached. You say different words to me."

I sigh, "She is killing me, mostly. If I can offer some fresh advice that I've just learned, even better. But I like when we talk like this, so I can stay current on what's going on in your world."

"Well, I hardly think you have the same problems at your WL job that I do at a water park."

"You'd be surprised."

Reflections

Even in seemingly hopeless situations, there is something a leader can do to improve it.

The same problem repeating itself in an organization becomes culture.

Growing Leaders

Every month I have a breakfast meeting with Karen, an employee on Vijay's team. Karen and I worked together a while ago, and rekindled our friendship when she learned I was working with the agile transformation. She has experience as a scrum master at another company, and really enjoyed working on an agile team. She was the first person I called after being asked to lead the agile transformation, and she was thrilled to be a part of the loosely-formed transformation team.

"I'm glad you were still willing to meet today, even though we are talking every day with the transformation work." I put my phone away and settle in to my breakfast.

"Of course, Joel. Based on the changes we need to make for the transformation team, I thought today would be a can't-miss episode."

I chuckle, "I have a feeling they will all be that way indefinitely."

"You didn't show it at all in yesterday's meeting, but that whole thing with Thad must drive you nuts."

I smile back at Karen.

"Yeah, that's about all you can do in public, isn't it?"

I keep smiling.

"He is such a jerk. But I'm thinking about the people he's working with. They probably don't have issues with you or agile. Probably only with Thad, but that's not our problem. So I was thinking about what you said about not being defensive, that we are going to launch ourselves the same way we would if the operations area wasn't going off without us."

"Yes. Any feedback for me?"

"Well, yeah, it was a good message. Having a strategic offense and all that. Agile is grounded in collaboration, yadda yadda."

Karen is unafraid to literally step past Thad to build a team that is all things continuous learning, not just agile, not just lean. She points out that if we work to become agile and get better at it, we will need different methodologies and thinking. DevOps, for example. And, the way technology changes so fast, there are likely going to be methodologies available to us that don't currently exist. We are shortsighted if we think agile and lean are the only ways to transform WL. We have to be open to any possibilities, if we are serious about change.

"This strategy doesn't undermine Thad, because he's purely focused on lean."

I begin thinking aloud, "Our team could be cross trained. We could have specialists in each methodology, but everyone on the team would be a generalist."

"Right, Joel. You just sent me to a bunch of training; I have an idea of what was valuable for us right now, and what might be better for the future. I'm sure you have some ideas too, but I would like to help with that."

"You're onto something, Karen." I wait for her to tell me more.

"I also think with cross training, we will be a stronger team because of our knowledge, and our ability to step in and help when needed."

It's so obvious to me now: Karen is more than a key player of this transformation. She is an enthusiastic leader of it.

"Karen, I need someone to lead this cross functional team. Someone who can help carry forward the vision; someone willing to take the heat and the politics. There will be a lot of that," I pause.

"That's you, Joel."

"Yes, but I can't be as involved with the team as you want. Not because I don't care, but because I can't do the local leadership, and the sponsor and stakeholder work that this transformation will demand."

Karen nods, "You will be constantly talking to others about what we're doing, and how we can help them change the way they think about their work. Some people will get it, and some won't."

"The leaders of WL probably don't think this work is for them; just for their teams. That's going to be a big part of my work. I can see it already. But, let's get back to the topic. Are you interested?"

Karen nods yes. She tells me her first needs for the team, other than a name, is a project manager, project assistant, and a people leader. We will also need some team members, and she has a few ideas. We will also have to hire some team coaches to help us help others.

"You're the people leader, the team leader, Karen."

"I don't have much experience with that."

"I need a servant leader. Someone who can remove impediments, see the flow of the work, and there will be a ton of it, and relentlessly share the vision of where we are going with the team. We need to change WL, and that might not be possible without changing the way we lead. I need an open mind like yours, Karen, to show others how to do it."

"Scrum master…" Karen is thinking.

"You'll have some performance responsibilities, too," I offer. "I have asked Vijay to help you get this team rolling. I asked him because first, he's on my team, and I can easily off-load some of his responsibilities to free up his time. Second, Vijay has lots of experience building teams at WL. He knows all the ins and outs of HR and budgets."

"Yeah, we will need a strong team that can take the heat, and still work well together…"

"Yes," I wait. "And, of course, there is a need for it right now. Not a rush, but there is urgency."

Karen smiles. "You want me to be the new kind of people leader. One who has the qualities and competencies of a scrum master, and has people leader responsibilities, too. One who's willing to step forward and mess up, because that's how we learn what works and what doesn't."

"Yes. And, you can do this because you will be grounded in the vision of the transformation. Vijay will be right there with you to get you started. And, you know I have your back. It's going to be very difficult work, and very exciting work."

Karen laughs, "Yeah, you can't promise much. But you can guarantee excitement."

"This work is going to turn WL on its head. That's why I believe you are perfect for the job. It will be exhilarating, hopeless, and open for interpretation. We will be popular and unpopular; loved and hated. You being a leader not cut from the corporate cloth may stir up trouble in the future that we can't even imagine today."

"I'm not perfect, but I am up to the challenge, Joel. Thank you for the opportunity."

"I want to help you build the team, but I want you to have autonomy with the team. You make hiring decisions, what kind of training we get for the team, who we hire to coach us, all that."

"You sure you don't want to be involved?" Karen looks uncomfortable and excited. It's a look I hope to see on many other faces in the future. "You're telling me to GO. And so I will do that."

"I don't know much about the how of this transformation work. But I believe that I can't control it. I believe it's best for the transformation to have solid sponsorship and then strong local leadership to keep the team engaged, learning, and hopefully, excited about their work. With you as a leader at the team level, you can be there for them every day, and still be able to connect with me."

"How does your assistant Marilyn play into all of this? She's so good, so much more talented than just being an awesome assistant for your team."

"She's along for the ride, for sure. But I don't want to answer on her behalf. Talk with her. See if she wants to join in the fun beyond what she's already got going."

"I might just do that, Joel."

Karen's thoughts are racing. I can just feel it.

"What about Vijay? I'm wondering what's in it for him? He's not just helping you with this team. That won't be very fun for him. And, he's losing me as one of his team members."

I assure Karen that Vijay will have plenty to do with helping lead the transformation. Talk with him about it. Not to seek permission. Collaborate openly and honestly. Ground yourselves in our vision, and know I am here to help anytime. Working with him to get the team rolling will be a full-time job. Once that work slows down, Vijay will begin more work in helping lead the transformation. That work is wide open at this point, and Vijay is just the kind of person to take on the challenge.

Karen smiles. "I love it. But I have a few demands beyond the first ones I mentioned."

"Demand away. What do you have?" I smile.

"I want the team to sit together, and I want to sit with them, not in an office. It's called a collocated team." Karen describes the scrum team setting she used to lead, and how powerful that was for the team to sit together. "Decisions were made faster, and more work got done."

"So, you need some facilities help."

"And people need to be 100 percent dedicated to the team. None of this 20 percent crap."

"Vijay can help you."

"And I need a bunch of money. A budget."

"*Investment*, Karen."

"That's right, just like we discussed in our team meeting yesterday. We are investing in changing WL."

"Right, and your team will get a chunk of the investment to lead the transformation. We will meet often to review how we're spending our investment. I'm sure it will seem huge as we begin, but it should stabilize over time. Rick gave us a healthy amount; he believes in giving our work the best start and support that WL can offer."

"This might sound forward," Karen shifts in her seat. "But you did ask what I needed to lead the team. Like you said, our team and you will have a daily standup to sync our work. And, you should probably sit with me and the team at least once a week. You know, go and see the work."

"It sounds forward for WL, but not for how I need us to work together, Karen. You and I have to commit to being honest, and saying what's on our minds *when* it's on our minds. We will have to use care with the rest of WL, but here, and within our team, we must practice what we are selling," I smile. "And yes, I will absolutely spend at least one day with your team. Our team."

"You and I should spend time together, too, Joel. You'll need it to relentlessly share your vision with me," Karen winks. "There will probably be some really dark days ahead. I will need to be constantly reminded of where we are going and why we are doing this."

"You got it, Karen. I don't see any other way that this will work. This is my new job, we are a new team, so I will be here for you. I still have other responsibilities, but I will learn over time how that needs to change."

"Can you please tell me those strategy items again? We just talked about them in our meeting, but they are suddenly more real now."

"Sure." I open my notebook and slide it across the table to Karen. "Some of them are more like tasks, but they were identified as key to getting us up and running."

She copies the list on her iPad:

> Who goes to Essence of Scrum?
> Who gets certified in scrum?
> Business partner involvement.
> Hire coaches for the teams.
> Promote the business value of agile and the wins.
> Get expert organizational change help.
> Align with Thad.
> Determine the first environment change.

"Joel, these things will be the first activities we do." Karen leans across the table. "Are you sure you are willing to let go of this long list of *whats*, and trust us with the *how*?"

"Yes, I am sure, Karen. If I start to cramp you and your newly forming team's style, tell me. Just be direct and remind me of *what versus how*, and I will back off."

Karen smiles, "You're really serious about all of this. I can't believe WL is doing this, and that you're asking me to join in the fun. I'm so stoked about it, Joel."

"It's going to be a real undulating journey, Karen."

"Just so we're clear…I want to be a leader, not a manager."

I smile. She is going to be great.

"It's going to be difficult, Karen, because we don't have many leaders at WL. They haven't been allowed to be leaders, for the most part. Managers have had the authority here, but the actual *leaders* here are the ones who take responsibility."

"Right. You really get it, and I am glad."

Karen has been studying leadership as it relates to leading change. She also studies it from the agile and lean perspective, but the view of the overall change is what interests her most. She believes we are going to have to sell this change to people through our leaders being leaders. Not leaders being managers, where the change is forced on us, or told to us in a factual way. We do need some people to be managers, but in the case of leading the change to agile and lean, we need leaders.

I wasn't sure about the factual part, so I asked. First, Karen assured me we won't have a spun-up, don't-say-the-word-lean strategy to get people to follow us. And, isn't that what we're after, getting followers?

"Joel, it's really tempting to think that we can just tell people about agile and lean, and they will buy into it. You know, that they will start…becoming. Maybe some sheep will, but not most of us. And, it's tempting to think that we can logically sell agile and lean to people. You know, tell them it's good for them, they will like it, and the outcomes are awesome for the company."

"Ah, I get it. Great point. We are so cerebral about everything. But this change is different. It needs connection…to their work and then to the customer."

I like where she's going with this.

Karen hesitates for a moment, "I…I have one more thing…"

"Go ahead."

"I didn't plan on bringing up this idea today. I thought it would take a while for you to give our team the go ahead on things. So I was just going to sit on it until the right time. I realize now I don't need to do that anymore."

I nod in encouragement.

"I have an idea to get our transformation off to a healthy start. We will have a vision and all that, but I'm talking about what we do with the vision. How do we carry out the strategy to make the transformation happen? We know the odds are stacked against our transformation, so it's going to be very difficult. On top of that, things are getting a little weirder with Thad and his personal agenda. So we need a bullet proof strategy."

"I agree, and I love it. What do you have in mind?"

"Well, I saw a video on transformations from this consulting firm called AB² Consulting Services. They used a really easy formula to guide how to think about a transformation. Super high-level, but I think it's something we need." Karen chuckles. "Almost like a mindset for the mindset change."

Karen opens her tablet and shows me a picture of a gecko. She laughs, and scrolls down to the content, and then shares it with me:

Think GEKO

1. **G**et a healthy start. Learn how to get the right people in the room, and how to objectively look at the work.

2. **E**xercise to see the work. An exercise with powerful questions designed to help you and your stakeholders have objective, productive discussions.

3. **K**ick it off right. Don't keep it a secret, get the training, and relentlessly share the vision.

4. **O**ptimize through PDCA. Get feedback, and learn from it. Help others to learn, too.

"Huh," I say. "It is simple. If you think it's worth a try, let's do it."

Karen smiles, "You don't have to give me an answer today, but let me know. I see us using this to get ourselves up and running, and then moving on to another strategy to get us through the next phase."

"Do you have to buy it from them?" My head starts wandering off to my concern about keeping our methods and tools narrow, so we can remain focused.

"No, this is free. But they do offer consulting and coaching."

"Okay," I sigh. "I'm getting into the *how* of things, Karen. Your idea has business value, and we should go for it. Our startup needs all of the help it can get. I rely on you to fill in what each of these four points mean."

"So, in other words, I can run with it."

"Yes, please. I trust you, and can't wait to see what you do with this. And, I'm here to help when you need it."

I sound confident in Karen, and energized by her initiative, but I'm overwhelmed. How is all of this going to work? My head hurts just asking that question. I currently lead a team. They are great, but they need me. When Rick offered me this role, he said he would give me what I needed to be successful. I got an investment of money and Thad. But I also got the freedom to develop the vision for the transformation. So, I have everything I need.

But I don't feel ready. I am ready, but I don't feel it at all. That's the real stress. Eve would ask me five whys about that statement, to really unpack it. Not why I don't feel ready. The reasons for that are fairly simple: I've never done this before, and I am not an expert in the methodologies I'm introducing to WL. Something, or many things, are holding me back from being courageous about this work. What holds me back from having the courage to take the next step forward?

I'm definitely worried. Eve would ask me the more abstract: why am I *worried* about not feeling ready? Well, if I'm not ready, I might make bad decisions that impact the entire organization. It's not all up to me, but I could really mess it up.

Then there is Thad. He is a huge problem that is probably going to get worse. If we are a mess, it could make the entire transformation effort lose credibility. As my daughter Caroline would say, we as a team and as leaders won't look *legit*. Without followers, we won't be successful.

Yet even if I take the Thad problem out of the situation, there is more. It's Rick and the people above Rick. Not their personalities or work styles, but their belief in change. They say they want to change, but are they serious? They say they are accountable, but mostly at WL, that just means they might get called on the carpet, but they won't experience any actual consequences. The teams and people below them will. They are the expendable *resources* anyway. It's a messed up system. I'm sure Eve has a name for this, and a deeper understanding of it than I currently do.

The stress, the worry, is that I will be the one to find out if they are serious. Both the transformation team and I will have our careers on the line to find out how much change can really happen at WL. That's it. That's the crux of the problem. I'm betting my job that WL is willing to transform into an agile and lean learning organization.

The thing is, no one can ever know the answer to that question ahead of time. Not for this transformation, or any other. And no one knows how it will play out either. Not knowing how far WL will go to change, and how it will actually change is the opposite of any work I've ever done at WL. My past work has been linear: defining a goal, then a strategy, and then driving for it at all costs. If things aren't going well, get more people on board. Or maybe, get rid of some of the people who are thought to be holding the work back from success.

To be courageous today, I will have to push past the uncertainty, and take the next step because of who I am. I have to take the next step because I am an agile leader.

Reflections

I need to remember I have the *what*, not the *how*.

Karen will be a true leader of our transformation, but still needs my connection and support.

Invest, not budget: we do valuable things in which we invest.

Accountability at WL is questionable at layers above me.

I'm betting my job that WL is willing to transform into an agile and lean learning organization.

Facilities

I'm packing my laptop to leave for the day when Thad pops his head in my office.

"Hey Joel. Got a minute?" He's already in my office, so even if I don't think I have a minute, I have one.

"About to head to my son's baseball game, but I have a minute."

"There is some pushback from facilities about visual management. I'm hoping you can take care of it with them."

"What sort of pushback?" I'm now ready to leave, but this could be interesting.

"The kind transformation teams don't like," Thad grins and at the same time sticks his tongue out for a second. Those intense, blue eyes and his tongue sticking out, he really looks like a reptile of some kind.

"Joel, not everyone is going to understand how we work. As we drive forward with change, we are going to create waves. We need to make sure people come along with us. That's a real important thing to learn, here." Thad is all-knowing, and I am fortunate to be educated by him.

"I need you to talk with the facilities people. You should get a meeting invitation from them soon. Well, we will both be invited, but I'm not sure I need to attend this one because I have so much on my plate. It's like Thanksgiving dinner, Joel. I don't know where to begin," Thad sticks his tongue out again. Wow, what is up with that? This new facial tic or nervous habit is disturbing, and very distracting. I'm used to Thad looking at his laptop, at the ceiling, out the window, at his phone, anywhere but at me. Now, this new stick-out-his-tongue-thing he's doing is making our interaction very uncomfortable. But I got this. I'm not going to be a victim of Joel's social awkwardness.

"Uh, I'm happy to talk with facilities. Do you have any other context to offer?"

Thad shakes his head no. "There isn't a lot of detail at the moment. Know that this is big change for them. You need to make it clear that this change is happening, with or without them."

"Has Rick or any other senior leader come to you with this feedback? In other words, anyone above us involved in this?" Besides not trusting Thad at all, I need to learn if there is any other pressure on him beyond facilities. So if I can, I want to avoid any surprises or misfires.

Thad assures me no one else is *in the mix*; the facilities people approached him directly about their concerns. In fact, he knew how much I wanted for us to be aligned in our work, that he saw this situation as an opportunity. He thought this was a great way for us to begin working together. Facilities wanted to dive right into the problem with him, but Thad asked them to hold off. He told them that I should probably be involved since the tech teams I'm working with will likely cause the most change.

"Now, Joel, you and I both know that I could have taken the bull by the horns with this one and just solved it. I would have gotten facilities to give us everything we needed, and made it feel like it was all their idea," he makes his lizard face again. "But I didn't because taking on this problem will be great practice for you, Joel. And also, it will get you some exposure with this group. I want to do that for you. Besides, facilities has probably been off your radar."

Huh. I need practice. I need exposure. And Thad, the all-powerful new guy, can give me that.

"I've laid the groundwork with them, but it's still going to be difficult to stand firm, Joel. But I know you have it in you." Thad is attempting miserably to encourage me.

There is so much happening here, and as a result, I'm speechless. First, his idea of alignment is to give me assignments. He thinks he is overseeing me and the transformation, and delegating work as needed, or when I need a growth opportunity. So, when I get work like this, it's because Thad is doing me a giant favor. The worst part is that he's using my desire for us to be aligned as a way to get me to buy in to something. It was easy to pick out because Lora, our former VP, did it all the time. Lora and Thad probably received their manipulation training from the same school. Yet Rick hired both of them. He's not a manipulator; he can't possibly like this quality in others…

Thad believes he is a dictator of the transformation. He only wants us to go in and tell them how it is. I don't hear him saying anything about wanting to understand more of the facilities department's needs. Without talking with this facilities group at all, he believes he knows exactly what they want. He also knows exactly how to fix it, but he's not going to, because he's beyond it, and I need practice. He has a hammer of authority and methodology, and facilities is a nail. I'm also feeling like a nail.

Maybe it's not such a bad idea for me to be assigned this work. If he's going to act like a cowboy or a bull in a china shop, Thad could ruin things for the rest of WL. It's hard to believe one person could have that kind of effect, but Thad's positional power as a VP makes that entirely possible.

Thad was hired for his strong lean manufacturing background. He used to be a director at HGoods, where he led process improvements as six sigma specialist. Meryll, Jack, and I questioned how a guy with strong manufacturing process improvement could be effective at a tech company. It didn't make sense to us, until we learned how Rick found Thad. They met on a business flight, and built a friendship over the course of a year.

Lora had the VP job before him, and her leadership was miserable; she was controlling, manipulative, and used intimidation to get her way. Her technology background wasn't anything impressive or highly skilled, either. She made her way to the top by delivering and driving for results. No one cared that there was a trail of dead bodies behind her projects. So when she was fired, Meryll, Jack, and I were hopeful that we'd either get a better person for the job, or WL would not fill the job at all. Then, Thad was hired as the new VP. We were hopeful he would be a better leader than Lora, even just slightly better would be great.

We tried to give him a chance, despite all the red flags and social awkwardness. It was very clear early on that he was just a slight variation of Lora. But we couldn't help but think how realistic it was for a manufacturing leader to be successful leading a tech transformation. The real world experiences had to be very different. Tech employees are not factory workers on assembly lines. Well, there is some production element to their work, and it's a common complaint right now that WL is too production-oriented instead of looking at customer needs, and the intent of our products. Eve has been coaching me on this very topic, and how agile can create a more collaborative, creative environment. When persistent teams are able to focus on the work together, they are more creative, driving innovation and new answers to problems.

So if we are trying to get away from a siloed, production environment, why do we have a leader from a production environment? Because the CIO wanted comfort in making a big decision, and hired a friend to help him. Of course the bigger problem is how we're siloed in our quest to de-silo the organization. Ugh.

Back to talking with facilities, I don't have all the answers on the right way to handle things, but being aggressive with a new business partner is not how I want to work. Not because I'm afraid of them, but because the transformation team needs to build a solid relationship with them. I have a feeling our work together is just getting started. Why start it off on the wrong foot? Why not approach this work as a partnership? Coming in to the problem pushing them around isn't going to yield anything but pain. And, an early dislike of agile, lean, or any improvement efforts.

Huh. That's probably another reason agile fails. At least, at WL. Dictating change to people and teams, instead of…well, I'm not sure. There are probably many ways to do it right. One way I can think of is seeking to understand people and their work. For my leadership and relationship style, this feels like a solid first step. This approach gives people a chance to understand what's going on, and make a connection to their work.

Back to Thad, the best choice is to do what he's asking without making it a big deal. He said he's laid the groundwork, whatever *that* means. And hopefully, the people he's already talked with aren't too annoyed by his *groundwork*.

"I will be happy to meet with them, and I will have them add your name to the meeting as well. This situation affects both of us."

"Sounds good, Joel. I know you will build some good inroads with that team."

He has no intention of joining me for these discussions. For this meeting, or likely anything else.

"Thanks for coming, everyone." Lucy, a manager from facilities management, smiles weakly before opening her notebook. "The purpose of our meeting is to discuss facilities policies, and teams' adherence to them."

I'm at this meeting along with my peers Ben and Dipti, as well as WL's first scrum master Waylon, and Vijay, a manager on my team. Just as Thad had said, we were pulled together by facilities to discuss the type of work our teams are doing.

The real reason is that facilities is pissed that people are making their work visible: teams are putting their Kanban boards and other items on the walls, and even on windows and backs of doors. They are using giant, white sticky notes (flip chart paper that has a sticky band across the top) and small, colored sticky notes. Facilities said it looks bad, and that senior management also thinks it looks bad.

The glue on the sticky notes doesn't ruin the paint. Yet I can see how radical this must look to others. But I also know that the logical place for the teams to keep work like this visible is…where they work.

Lucy looks around the table at us, and squints her eyes at Waylon. "I'm sorry, I don't think I know you,"

If I didn't know better, I'd guess Lucy was from HR, not facilities. Her mannerisms were so classic WL HR. The lingo, the squinting eyes, and projecting the feeling that we are guests in their royal domain. Then again, I've never really worked with facilities, so I guess I should stop imagining I have. *Keep an open mind. Seek to understand.*

"Hi," Waylon holds out his hand across the table. "I'm Waylon."

Lucy has no choice but to shake it. She's not hostile, but she looks like the hand shaking thing is causing her mild pain.

"I'm the scrum master for the team *Can't Make This Up.*"

Lucy turns her head to one side with a small smile. "I'm not sure you're in the right meeting, Waylon."

"He is," Ben says. "*Can't Make This Up* is one of the first ever scrum teams at WL, and Waylon is their leader."

Now Lucy and Ben, the two leaders, are talking as if Waylon isn't there. He sees what's happening, and with a slight flash of amusement on his face, he lets them talk it out. I'm surprised Ben is doing this to Waylon. I think it proves that old habits are hard to break, even with the best intentions. When we are pressured or stressed, we go for comfort.

Lucy points to Ben, "Oh. So he's in your area. With that team with all the sticky notes."

"That's right," Waylon speaks again, which mildly startles Lucy. She is accustomed to leaders speaking on the employee's behalf. So, when the person actually doing the work is in the room, and she has to talk about them, it's uncomfortable, because it is real. Waylon is real.

At this moment, Ben realizes he shouldn't have spoken on Waylon's behalf, and stops talking. He gives Waylon a look of encouragement.

"I lead a scrum team that is trying some new ways of looking at their work."

"So, a scrum master is a leader?" Lucy asks, and looks at Ben. "I thought you were the leader."

Waylon explains to Lucy the role of a scrum master with as few details as possible. He does so in a way that makes this awkward situation feel better. He's a great ambassador for scrum and for his team. I can see why he has the job, and I'm so very glad he's in the room.

Lucy clears her throat and leads the charge, "Okay, then. We have noticed many teams suddenly violating the policies we have in place about our windows and walls."

Lucy tells us that there are strict rules about keeping our walls and windows clean. It's actually called the Windows & Walls Policy. Effective immediately, we as leaders are being asked to make the teams stop using them, and contain their work. Oh boy. We probably should have talked to facilities before these first scrum teams started. In fact, we probably should have talked to facilities when Vijay and his team put up their Kanban board.

Although Lucy is telling us we need to stop using the walls and windows, we may just need to share more about what we're doing, and apologize for not engaging facilities up front. Meetings like this happen often at WL. It's the why-weren't-we-told-about-this meeting. The person or team calls the perceived violators together for a meeting. The problem is discussed, sometimes at great length. Usually, the group arrives at an accord: stay in communication with each other on the changes, and don't work in a silo. A committee is sometimes formed. Or, if there is already a committee in place, as an olive branch, the person who was wronged is invited to join the committee. Most of the time, no one has any ambition to join another committee. They just want to be informed.

Unfortunately, Dipti didn't see the meeting pattern, and before anyone else can answer Lucy, she dives into the weeds on the problem. "What exactly is happening with the windows and walls? Are they getting damaged? Was someone injured?"

Lucy holds her weak smile, "Nothing like that. Yet that is exactly why we have these policies…" Lucy starts tutoring us in the vision and mission of the facilities presence at WL. It's actually impressive that she knows her work impacts to the company at the strategic level. There are not many people who understand their department's role within WL like this. She has no practical application of it, but she knows it.

Ben gently apologizes for not having facilities at the table when setting up the new team. Lucy smiles back at Ben, and shares that her team was *shocked* at the way this new team was working, and that *they didn't know what to think*.

Ben carefully pressed forward, wisely avoiding any words about the transformation in his apology. He shares with Lucy that this team is an experiment, and there is a special project, and working in a different way to deliver it that requires a high degree of collaboration, and a way to make their work visible. He notes Vijay's teams are doing similar work.

"So, Waylon's team and Vijay's team are the same team?" Lucy asks. She knows the answer is no, but makes us work for it.

"They are two separate teams," Vijay smiles.

"So now there are two teams like this that we didn't know about?" Lucy sighs, acting like this is the first she's heard of the facts, even though she knew about Vijay's team and invited him to the meeting. This is not Lucy being mean, this is Lucy trying to get a grip on a problem without speaking to us in a direct manner. WL culture doesn't allow for it, even though all of us desperately need it.

Vijay continues with the apology Ben started. Then, he shares with her how excited the teams are to have this work. He describes how they love the efficiency of working together in the same space, and meeting at the board to talk about their work. This point seems to get Lucy's attention. That makes sense, because part of her mission or vison words included creating an environment of positive energy that contributes to the new vision of *grow happy, grow strong.*

As the meeting progresses, Lucy's original request of asking us to stop our teams from working this way is on the back burner. We are discussing the details of a collocated team, and Kanban. Lucy is invited to each of the teams' areas to interview the team and see how they work differently in their space. She said she would have to check her calendar, but would like to take us up on the offer.

Vijay shares the names of other companies with teams like this, and offers her ways to get more details about how these teams work. He wisely doesn't tell her that teams have been working this way for decades, and that WL is behind in how its people are asked to work. Vijay encourages Lucy to share with her other team members as well, because maybe some of them are familiar with this different way of working. We may get different ideas from them that we don't have. This statement throws off Lucy; she's not used to people considering that they can learn from her team. She's not used to collaboration. It seems to really brighten her disposition.

Lucy sighs, and takes off her glasses, "Can we please go off the record for a moment?"

"Sure," I say. The room is getting thick.

"Are you all working with Tad?" She asks.

"It's Thad," Dipti corrects.

"Oh, sorry."

How to answer that question? I have to say yes, so I do. Lucy validates again that we are off the record, and I assure her we are. Ben, Vijay, and Waylon nod, too.

"I'm sorry I began this meeting on the defense. It's not my usual way of working. It's just that…when Thad contacted me, he was so forceful." Lucy looks to the ceiling. "Actually, he was a jerk."

It was tempting to say something at this moment, but the best thing was to let Lucy talk. She had more on her mind and it's obviously going to impact how we work together. Or don't.

"You see, I have a lean six sigma background from my prior job. I don't get to use it much where I am right now, but with this new change coming, I am excited for it. I think there are a lot of opportunities to make WL a better place to work."

"That's great, Lucy," I say, and then apologize. I ask what happened.

"He told me that there is big change coming, and that we had to get on board. He said that we are doing process improvements that will require facilities to change the way we work. So I asked him if we were going lean, and he said no, that we were going to use a new, exciting tool called a rapid improvement event."

Lucy grins, "I didn't know whether to laugh in his face or begin an argument. So I held back telling him about my lean six sigma skills, and asked for more details.

Thad told Lucy that the new rapid improvement event tool is used to improve how people work by refining processes and looking for waste. Then, he asked if she could keep a secret, so of course she said yes. Then Thad informed Lucy that the rapid improvement events will eliminate unnecessary work, in the form of jobs, which will save WL tons of money. Our expenses are too high and this is a way to do something about them. He apparently gushed about how we are working in exciting times, and that we are going to do the most amazing things with less resources.

We are shocked by the story, and by Lucy's willingness to share it.

"So I asked him a bunch of questions about the rapid improvement event tool, like how long has it been around, will everyone at WL have to use it, and when will we begin using it." Lucy now had a smirk. "He tells me it's a relatively new tool, that everyone at WL will have to use it, and that the plans are being kept *close to the vest*."

Thad said he had some ideas out of the gate on how to improve the facilities group. He talked about bottlenecks and unmotivated teams, and no one being able to solve the problem, because they haven't had any such training. Then Thad began talking about the assistant director role, and how rapid improvement events would show that facilities doesn't need that role anymore. Too much work gets stopped there, and can't flow to the customer. By eliminating that role, we push decision making down to the people, and save money in the process. Even though this infuriated her, Lucy didn't bother telling Thad that the facilities area didn't have any assistant director roles. She couldn't believe a VP was talking this way to her about change, or about anything, for that matter.

She was also very concerned that this was the direction WL was taking. Just when it felt like we were on the cusp of changing things for the better, this seemed so heavy handed and dictated. Yet here was a VP using a full command-and-control press on her.

Then Thad told Lucy that she needed to get her facilities team on board *right away*, so that we can help be part of the change. He said that the areas that are slow to change will be the most vulnerable to getting their jobs evaluated for waste. Lucy said this was the point where she had enough of his crap, VP or not, and called him on it. She said she knew it was a risk, but if WL was going to turn into the kind of company Thad was leading, she didn't care if she got fired.

"So I tell him I have a lean six sigma certification, and that he doesn't have his facts straight about our department," Lucy laughs. "And you know, he didn't believe that I'm a black belt. He started quizzing me on what is

DMAIC, what is *muda*, but I didn't take the bait. I wasn't going to put up with his BS."

"So what did you do?" I'm rapt with her story. "And, later, please teach us about DMAIC."

"I offered to have him come to my office to see my designations, and he backed off. But he didn't *stop*. He kept pushing. He was saying we could have a great mutual relationship where we help him and his team with the teams transforming, and they will help us get rid of the waste and the extra resources. He actually said that our resources are holding us back from our full potential."

Lucy still has that smirk when she tells us how Thad has this odd mannerism of sticking his tongue out every once in a while. She believes he thinks he is being personable with it, but when he does it so often, it's just creepy. Lucy told Thad that she would be happy to talk with her department about helping his team, but they did not need his team's help with looking at their processes. In fact, she had already helped streamline many operations in her department. Thad urged her to reconsider because her experience was likely far more limited than his or his team's, and wouldn't yield the best results.

Vijay is taking notes, and Waylon is listening attentively. Ben is ever-so-slightly squirming in his seat. I'm just taking it all in. All of it is so unbelievable, and yet I can absolutely picture Thad acting as Lucy describes. Thad's sales pitch and leadership was appalling, and Lucy's strategic responses to his intimidation were terrific. Serendipitous, in fact. What would be the odds that the first place Thad looked to push around would have a representative who was a lean six sigma black belt with some confidence in herself? Incredible odds.

Lucy told Thad that her experience was deep, and her team was full of great people who contribute value to WL every single day. The fact that WL is now embracing process improvement is exciting, because now they will have funding for changes that were turned down in the past. In grand fashion,

Thad offered one, last time to help her area with rapid improvement events, so that their team's proposals had less waste in them. This way, they may be more successful making the cut for corporate planning and prioritization. Lucy said she had to shut down the meeting before she lost it, so she told Thad her team would be happy to meet with his team about helping change the work environment for these pioneer teams.

I apologize to Lucy again, and state that as this transformation gets rolling, it's going to be a little rocky. There are some different ideas on the approaches to changing the way we work. I'm making it my personal goal to unify our efforts so that we offer clarity and vision to people.

"He's a nut job, Joel," Lucy says flatly. "I feel bad that you all have to work with him."

Okay, so I can cut through the crap with her. Vijay's eyes are wide, and Ben is paralyzed. In fear, I think. They haven't worked as much with Thad as I have, so this entire conversation is probably startling them. When I glance at Waylon, he has his poker face on. This is not going to be easy to lead through without making Thad look bad. Well, he's done that all on his own. It's going to be difficult to focus on leading the transformation with Thad wreaking havoc ahead of, or next to, me.

Right now, the leadership challenge for me has two parts: first, how to lead through the after-effects of this meeting. I have a director and a manager both shocked by a VP's behavior. Second, how in the world am I going to approach Thad about this problem? He's breaking so many rules of basic leadership, and in addition, it sounds like he's bending a methodology to his liking, and hoping no one notices. I didn't predict this moment would happen, but I was concerned about Thad's style and that he would try to push his way over and on top of people and areas.

My last attempt at trying to get him to see things beyond his way didn't go very well.

But Lucy is my focus right now. I thank her for her candid story, especially since we didn't know each other before now. I express my hope that we can partner on the facilities work, even though we got off to a rocky start. I tell her I'm impressed with her background and skills, and that if she is willing, we could learn from her. Lucy was energetically agreeable to both.

"But there's more, Joel."

Of course there is.

"Have you heard of C²?"

"Uh, yes I have," I say, knowing that Ben, Vijay, and Waylon probably haven't heard about the newly formed Culture Committee. "Everyone else here probably has not, because it's a small committee."

"Well, Thad told me that he is *spearheading* C², a new committee designed to change the culture at WL. And I never did learn what C² stands for. Anyway, he said our culture is going to undergo a dramatic change, and that facilities should be ready for it. Do you know anything about that? Do you think it's the same work as the stuff we're doing with your teams? He was such a jerk that I didn't have the stomach to ask him any questions about it."

Well, this is awkward. I tell her what C² is, and confirm that Thad is leading the work. I don't know what *the work* is going to be, because the group has no vision or mission. The group is waiting on Thad to attend one of our meetings to share his thoughts and those of the senior leaders who charged him with this committee.

I also share that I'm seeing some big connections between this transformation work and that of C². In fact, since my first C² meeting, I learned that culture changes is an outcome of environmental changes. That's counter to what C² is set up to do. Well, we don't have a mission or vision for C², but the overall idea of the committee is to come up with ways to change the culture.

Lucy chuckles, "So we have set up a committee that is the anti-pattern of the outcome. And the leader of the committee has yet to show up and tell the team what he's thinking and what senior leaders are thinking. That's great."

"How long have you been with WL?" I ask. "New committees pop up all the time."

"Three years," Lucy sighs. "Long enough to understand why C² is set up backwards."

Meanwhile, my three partners are still taking this all in.

"So C² is brand new and has no vision." I smile. "I wouldn't worry about it, and definitely don't plan anything around it."

"But there are people from all over the enterprise on this committee," Lucy is intent. "Surely many eyes are on it, and something will have to happen out of it."

"Lucy, you've been here long enough to know that we do love action. We have to do something if we are going to prove our worth. So I expect something will come out of it. I can't promise that facilities won't be affected, but most of the ideas I've seen are lighter than actual transformation work."

Lucy smiles, "Please don't tell me there will be a newsletter. No, wait," she feigns excitement, "a newsletter tied to a rewards program."

I stare blankly back at Lucy. It's all the answer she needed to begin giggling.

"I'm sorry, Joel. It's really not funny."

"You get this, Lucy. In spades."

"Okay, so Thad is really stirring the pot, and I think he enjoys it. Talking about job elimination, waste, and changing the culture all at once. I think he was trying to get me to buy-in, so I could be a hero for my department, and give him a feather in his cap. I can see through it, but others might not."

"If you can do your part to not fan the flame, we'd all appreciate that."

"Well it nearly derailed us from understanding how we can help each other," Lucy turns to Waylon and holds out her hand. "Waylon, I'm sorry I gave you a hard time." Lucy is now a different person than when the meeting began.

Waylon shakes her hand and smiles back, "No problem, Lucy. I'm glad we cleared the air on a few things. I feel good that we can get this out in the open. If we're going to work together amidst this other uh, madness, and succeed, we will have to be real honest. You can count on me for that."

"Thank you, Waylon. You have my commitment to being honest, too. And I'd love to learn more about the scrum master role," Lucy says. And the two of them schedule time for Lucy to come to where *Can't Make This Up* is located, so she can observe the team's work and gather their needs. Waylon is going to tell her about the scrum master role, and Lucy is going to give Waylon some lean background.

Then Lucy turns to me, "Joel DMAIC is an acronym representing the five steps of solving a problem. You can really geek out on it, just like any methodology. It's a cycle used to get rid of the defect or improve the opportunities in a process for business improvements. DMAIC means define, measure, analyze, improve and control."

I thank Lucy, and commit to researching DMAIC.

"Lucy, I know Thad got us off to a rocky start. I'd like to forget about that and move forward. I can't promise you won't be in an awkward situation again, but there is a lot of positive energy behind what we are doing. The teams Waylon, Ben and Vijay represent are showing signs of big success, and we are so pleased with them."

"So you want to replicate the model to other parts of WL," Lucy adds.

"Yes." Vijay is pulling out of his shocked state. "Our goal is quality over quantity. It's very tempting to push this out to a bunch of people, but we are more interested in a strong beginning rather than launching a bunch of teams. This way, we can learn from what works and what doesn't, make adjustments based on team feedback, and then adapt how we might set up the next teams." Vijay looks to me as if to validate what he just said. I nod to give him my acceptance of what he said.

"We don't want to go too slow or too fast," Vijay smiles. "But we don't really know more than that about the pace of change until we get rolling."

Lucy praised us for approaching our work with a learning mindset of sensing and adjusting from what we've experienced. Lucy said that she will help her team approach helping us in a similar way. With both teams having a learning mindset, we will be more creative, and have better success working together. No one will be stuck in failure if we can learn from it. Her words were slightly different, but the meaning was the same as when Eve coaches me about not having all-or-nothing thinking.

I pause. "I'm wondering if you would like to talk with Karen, the leader of our team, about joining it?"

I can't believe I just asked a lean six sigma expert to join the transformation. Eve would be so proud of me for stepping past my fear of lean and becoming a quality jerk, to see the value the methodology can bring to an organization.

Lucy is slow to respond. "So, this is a big enough effort that you're building an entire team to deliver it?"

"It's looking that way," I say, struggling to not add a bunch more thoughts on the topic that will steer us off course. It's difficult because I am very excited about our connection, and the transformation.

It's my first big transformation leader moment: I hope I get it right. I tell Lucy that WL has the vision to be a more adaptive, customer-focused company. Just like that saying, *what got you here won't get you there*, we can't do things the same way if we are going to meet this vision.

"I hesitate to tell you that everyone will have to change, because I don't know that for sure," I say. "But this is a pretty big goal, so we're going to make a team solely dedicated to helping people change how they work." I sure hope I said that right.

"I get that, Joel. Even if some people don't have to radically change the way they work, they will be affected indirectly by the people who do."

"Right. Well, that's our theory anyway. I don't have a fully detailed vision of who is on this transformation team, but after talking with you, I think we need your help. A lot of it."

"And you want me to be on the team? Not my boss?" Lucy's eyes have a new spark in them.

I smile, "Yes. We'd like you to join us. At least, talk with Karen about it."

"To do what?" Lucy suddenly writes a bunch of notes.

"Join us, so we don't keep pissing you off."

Lucy smiles.

"And more importantly, Lucy, we need your help creating the future."

"I am thrilled to help your team in any way I can. I'm honestly not sure I want to work that closely with Thad. I will consider it."

"What the hell is going on?" Ben huffs.

"What is C^2?" Vijay asks. "Does this have anything to do with the regulatory fines problem or how the competition blew us out of the water? I heard some rumors about that."

"Hold on," I put up a hand. We are huddled in a conference room after our meeting with Lucy. I felt it best for the four of us to have a debrief session right then. A lot was said about Thad's behavior; Vijay, Ben, Waylon need some closure on it. The point isn't to get them to take sides, but to be a leader, and connect with them.

"Look, this is the part of leading a transformation that I'm not familiar with," I say. "Which is all of it, actually. But I do know that this is a chance for me to reiterate our vision, and assure you that all of your leaders are 100 percent committed to the change. But our vision is only partially formed, and some of our leaders misunderstand what *100 percent committed* means. We are not in a place where I can reassure you of much, other than I am honestly committed to this change."

"Thanks, Joel. I do appreciate your words, but I thought Thad was working with you?" Ben asks. He is upset, and rightly so. He's the director of the area with the first scrum teams.

"He is. We are. But as you can see, we have different styles." It's a lame response, but it's the best I can do. I'm hoping Ben and the group can see what's happening.

"Come on, Joel. This is more than a *style* problem," Ben shakes his head. "I haven't worked with Thad, but I have yet to hear one good thing about him. I know it's inappropriate to say this, but I just don't understand why Rick picked him to work here. He's worse than Lora."

"It is more than a style problem," Vijay chimes in. "But what else can we do about it? If we let him go on, he's going to keep causing problems, and worse yet, we will look like we are undermining him. Leaders like this can't stand it when people call them out, even in the most constructive way. They consider it a personal attack. If we approach him with the problem, Lucy could get in serious trouble."

Waylon is watching the exchange, not saying a word. So I ask him what he thinks about the situation. I startled him a little with my question, because at WL, leaders rarely ask their employees' opinions. At least, when two directors and a manager are together…

"We got our hands full with Thad. Either way," Waylon touches my sleeve, "you my man, are screwed. Because guys like Thad get away with acting that way. I've seen it at other companies. People work around them, and make excuses for them. And the people who do speak up get nailed, or worse yet, fired. This shows the rest of the teams what happens when you mess with the psycho leader."

"What advice do you have for me?" I ask.

"This work ends up being about the relationships, more than the positional power people have. This drives a lot of people nuts." Waylon looks at each of us. "But in the end, most leaders get over it, because they really like what the teams are getting done. So if you can build really strong relationships with the people who matter to the work, things should work out. No one is safe from a dude like this, but the strong relationships will help protect your teams and stakeholders."

"What have you tried with leaders like this?" It can't hurt for me to ask.

Waylon rubs his forehead and pauses. "Man, this is bringing back old memories of my former company, for sure. You can exhaust yourself and your team if you *try* to make a strategy around this dude. Working on the defense becomes a lot of wasted energy. Your teams get focused on fighting the psycho leader instead of making the most of working in a new way."

Waylon tells us that the transformation leaders around the scrum teams focused on building their relationships with business partners and other stakeholders. They shared the vision over and over, shared successes, and asked for partnership to keep moving forward.

As Waylon talks, all of us are glued to his every word. This moment was special to me. We hired someone for their outside experience, and now we are actually asking him to share it with us, and we are listening. And, Ben and Vijay look like they care very much about the work we're doing and the people who are doing it. I have good people in my corner.

Waylon goes on, saying that the leaders met daily at first to share the progress made, and the problems encountered. They collaborated on how best to help the scrum teams, and how to keep the business partners connected. This work and constant alignment built trust with the stakeholders. So when a rogue leader would come to the stakeholders with a demand that was off the reservation so to speak, the stakeholders had enough context about what they were doing to sense something was wrong. They would take the concern back to the transformation leaders, and the topic was discussed in an open, honest way. The rogue leader was always invited to these discussions, but they rarely attended them.

The transformation leaders chose not to address the specific behavior problems with the rogue leader. They could have brought a bunch of grievances to senior leaders and to him, but they chose not to lead that way. They determined they could waste colossal amounts of time documenting and detailing all of the wrongs this leader committed, and hope that something will be done about the leader, or they could chose to work on their offense. It might not seem like a waste if the leader got fired, or at least removed from the work, but early on, this team made a choice that their energy was best spent focusing on leading the teams and the work. When they put all of their positive energy into helping the teams flourish, the rest fell into place. Some parts faster than others, but it all worked out.

They wanted to reinforce the new habit by not getting wrapped up in the old habit that was still happening. They chose to stop feeding the old habit, focusing instead on reinforcing the new habit.

"It's just what I learned," Waylon sighs. "We just can't let this dude bring us down. He's one bad apple in a barrel of perfect ones. And for right now, we only have one bad apple. Even though he says he has a team, he hasn't any minions that we know of."

"Thank you, Waylon." Ben seems to be less upset.

"Sure, but I'm not sure if my experience will help us in this case. I mean, every company is different."

"We are grateful you have shared yours with us," Vijay high fives Waylon. "You've given us hope that one person, no matter how destructive they are, can't bring down a transformation."

"That's right," the high fives continue. "We choose positivity and a good offense."

"One more thing," Vijay says. "About the rumor I heard."

"Yes, you said something about this having to do with the regulatory fines," I say.

"Joel, I was going to find you today to discuss this anyway. I'm sure it's not true, but as leaders, I want us to talk and have open conversation," Vijay sighs, and then begins telling us that a few of his employees came to him, concerned about another team. They had heard that because of the regulatory fines and how we missed big with our competition, there is going to be strategic expense management. The company is going to pull back on a bunch of initiatives and people on that team are going to get fired. But they don't want people to know about it, so it's going to be in a Trojan horse called the *grow happy, grow strong* strategy.

"So, you're saying the company is cooking up a veneer for the strategic expense management that's going to happen?" Ben asks.

"I'm saying that's what my team members heard," Vijay corrects him.

Vijay said he knows it sounds a little too manipulative for the way WL works, but most rumors have some truth to them. He said he thanked his team for being honest and sharing what they heard, and told them he didn't know what to say about it.

The team members said some managers outside of technology have been asked to submit spreadsheets with their employees' names, and skillsets, and certifications they have. They were asked by their senior leaders to do it, not

HR, and they had just 24 hours to complete it. No one was told why they had to do it, but one of the managers spoke up and asked why, and copied all the other managers who were asked to do it.

So then they were brought together on a call that was hosted by a senior leader. The managers were told it was only to evaluate where the knowledge and talent was, because they haven't done that in a long time. The managers were also asked not to tell their employees about it. This was a problem because many of the managers said they didn't know if their employees had some of the skills and certifications on the spreadsheet. They were told to make their best effort to find out, but it was very important that the employees didn't know about this request.

I can't believe this.

Vijay continues, "Then a manager noticed that the manager names were not on the list, and asked if they should add their names to the list. They were told no, that there would be a different list for them. Someone asked more about that, but the person leading the call quickly shut down the conversation. That made it feel like the answer was shared in error, and the call got weird after that. Someone asked if this work was tied to the new vision to *grow happy, grow strong*. I don't know for sure, but I think they sensed the group leading this work wasn't in line with the company vision. They are a smart group. They don't want to do something out of alignment. They want air cover."

"What was the answer?" Ben asked.

"My team said it was a politician's answer," Vijay chuckles. "The kind of answer that neither confirms nor denies the fact. It was enough to shut down any more questions about vision. But this is how I think the rumor part began. An urgent request without a real reason for it, and without tying it to our vision, people were on their own to understand why. The awkwardness on the call only made it more suspicious and ripe for rumors. Who wants to mess with their team's jobs when they have nothing to stand on? Who feels comfortable working in haste when there is no trust?"

We are all speechless. I pulled us in this room to get us aligned, and offer my leadership around Thad's behavior. How to get us on track again? *Get it all in the open. State my path. Stick to the facts.*

"Okay," I sigh, "so we have this nasty rumor, certain managers being asked to make lists of their employees, C², and the transformation work that includes Thad running around freaking people out."

"Don't forget *grow happy, grow strong*," Waylon says. "And the fines, and that lame Green product."

"This is rich," Ben sighs. "And disturbing. But are any of us surprised?"

None of us are.

Vijay thinks this work is not driven from the top of the company, but from somewhere or someone on the side. He felt it was too disorganized to be anything else. We all agreed that we could speculate all day long on this rumor, or persevere past it.

"I just wanted to make sure you didn't have some corporate directives lined up that we didn't know about, Joel. I know you wouldn't purposely withhold something, but this stuff happened really fast, and you may not see all the moving parts."

"I don't, Vijay. And you're right about moving parts. Separately, we can't see all of them. Together, we have a better chance. And, if there ever is something like this, I will probably be asked to keep my mouth shut, or lose my job," I say slowly. "This is uncomfortable for all of us. And it doesn't feel like that is going to ease up in the near future. I want you all to know that I won't break that kind of request, but I can still be honest with you."

Vijay nods, "I get it. You can tell me there is something you can't share."

"Right."

There is a short silence, as we absorb all that happened. Ben clears his throat, "Well, who knew our meeting with facilities was going to be so lively?"

"What's next, team?" Vijay asks with his usual enthusiasm.

My run that evening went by in a heartbeat. Not just because I'm tapering my workouts in anticipation of Ironman Boulder, but because I was reflecting on today's meeting with Lucy. And, our meeting after. It was a daunting and energizing experience. Just a few discussions, and a rumor that has some truth to it, and WL feels entirely unstable. Without hearing our vision throughout the noise and the change, we are left to speculate on what might be next.

But I think I'm ready to finalize my assignment from Eve. I'm ready to share my thoughts on why agile fails.

Reflections

Vision grounds us. Without it being relentlessly shared, an organization becomes unstable.

Forcing people to change never works.

Our first scrum master is a great ambassador.

Facilities may be our first environment change.

The Mouth

"I'm so glad we got to hook up before the race, Joel. Hopefully it won't be that windy on race day." The Mouth playfully punches my arm.

"Yeah. It was a good ride," I smile. And, I think, good practice in finding a place for the *annoying*.

Actually, I couldn't wait to finish our 40-mile route. And I was grateful for the wind, because we couldn't talk as much. Well, Tony talked almost nonstop, but I didn't have to hear most of it.

I knew our ride was going to be like this, and still I went on it. Because Tony would think I was afraid to ride with him if I didn't. I'm a grown man and yet this guy brings out the high school mindset in me. Now I can check the box, and move on with my Ironman. Since he's in the area, there is a high probability that The Mouth will want to train together after the race. Ugh, listening to that constant talking… My training coach would tell me to get The Mouth out of my head. *Find a virtual place for this guy and put him there.* Just because he showed up for coffee at the same time as I did doesn't mean he has to camp out in my head. My focus is much narrower; I'm racing an Ironman triathlon in a week.

All this talk about focus. I did research Tony's Ironman times online. He's a really good swimmer, but that's about it. Okay, so anyone who can complete an Ironman is doing something really special, but Tony's times aren't matching his hype. I can't say from my own experience, but I've read a ton on the subject. Aside from his great swim skills, Tony is a behind-the-middle-of-the-packer.

"You surprised me, Joel," Tony grins. "You were able to keep up in that wind. I mean, you drafted behind me a lot, so I was pulling the most watts, but you…" I get another punch in the arm. "You devil! You're so freaking ready for this race. Must be that coach you're working with."

"I wish you the best for your race," I demur. Anyone else, and I would tell them that I'm rather scared out of my mind by this goal, and I don't feel ready at all. But I'm not giving The Mouth anything about my feelings. And I'm sure as heck not going to talk about my coach with him.

The Mouth smiles and shakes his head, "Thanks, Joel. That's really great to hear. You know, I have plans for a PR. You know what that is, right?"

Man, what a moron. "Yeah, personal record. I'm going to have one."

The Mouth laughs, "You sure will, Joel! Nothing else to compare yourself to, so enjoy your pressure-free day. I'm real excited for you."

"Thanks, Tony."

Tony tells me again, how much he and his wife have trained for this race. He thought about getting a coach, but then decided he had enough training plans from friends to customize a plan for himself. On and on he went. All I could think about was how I don't want to see this guy on race day. Or his wife. But I don't know his wife, so that should all work out. She's probably a very pleasant person anyway. Listen to me, imagining what Tony's wife is like. This guy is full-on in my head.

"Will your coach be on the course? This race has so many coaches out there."

"Actually, my coach is racing, too. A few of us coached by him will be out there, but we're all at different levels, so we won't race together."

This sends Tony off into a tailspin about how he used to work with a coach, but he wasn't getting enough attention. Shocking. So Tony fired him, found another, and fired her. After going through four coaches, Tony decided he could do better on his own. I'm not going to ask him, but this guy is so confident in his talents; why would he think he needed a coach? Probably just wanted to do it because everyone else was. This guy definitely doesn't want to be left out of anything, and he would be willing to pay to be a cool kid.

"So, you want to meet up with my wife and I before the swim that morning? We can give you some last minute strategy tips."

"Thanks for the offer Tony, but I'm going to decline. I'm going to stick to my plan."

"So…you're really gonna be an Ironman." Tony states it more than asks it. "You're gonna want to quit a bunch of times. You're gonna have to stay strong, Joel."

I agree with him. Isn't that what you're supposed to do to neutralize a wild conversation?

Tony punches my arm again, "That's right. You stay focused, my man."

We part ways, and my shoulders go down about three inches. Why did I do that to myself, and a week before my race? I know better than this. It's just a race, but it's my huge goal, and I shouldn't do anything to take away from my focus.

Driving home, I let go of my mistake. As I do that, a new feeling washes over me. My heart rate goes up a bit, and I feel warm. I try to resist it, but there is no point. The competitive part of me is fully awake.

A small smile crosses my face. I have the sudden urge to kick The Mouth's ass on Sunday. Badly.

Reflections

I rode with The Mouth because of cheap curiosity; I wanted to see what kind of athlete he is.

Finding a place for *the annoying* is hard.

I can't wait to kick The Mouth's ass.

Vision is King

"So just a few days until go-time," Eve grins. "Ah, a newbie Ironman triathlete in the making."

"That's right," I feign confidence, when in reality I'm pretty wound up. Just hearing Eve saying the word *Ironman* gets my heart beating a little faster. I'm not scared; I'm excited. Well, I'm scared, but I'm far more excited than scared.

Eve has completed five Ironmans. I know little about her experiences, other than that her first race was her fastest. She says it wasn't her best race, and that I won't understand that until I do more than one. At the moment, I can't imagine another Ironman.

After several attempts to discuss her races, I've given up. Eve insists this is my time for leader coaching, and her goal is to deliver the best value to me. It's a style or quality that makes it great to have her for a coach: she doesn't talk about herself much. This time is all for me. Sometimes, it drives me a little crazy, because I'd like to know her better. But I get it. We have to stay focused. It used to make me uncomfortable, but now that I'm in the thick of things with this transformation, I need every minute I can get with her. At least, I think I do.

"Joel, are you with me?" Eve pulls me back to our coaching session.

I sigh, "I'm here, I'm here."

"We'll go easy on you today," Eve chuckles. "Aside from the fact that you're pretty distracted this week, we had a big session last time. I want to make sure it all sinks in.

I open my notebook to my reflections, and my answers to my assignment on *why agile fails*.

"I'm very distracted. The race, of course is a big part of that. Another thing is all the events happening this week. So much is in motion."

"And so it goes as your transformation gets wheels." Eve raises her matcha latte cup, and we clink with a positive note.

"You look concerned. Do you think any of the events that happened contribute to your homework assignment on why agile fails? Or, should we discuss them separately?" Eve's green eyes are steady. She already knows the answer.

"Yes, I am concerned, and yes, they are all mixed together," I sigh.

"So why does it fail, Joel? Give me your reason or behavior, and then tell me the outcome of it that leads to failure."

I look at my list:

Why Agile fails:

Forcing people to change never works.

Teams don't get the support they need.

When stakeholders don't get a single vision for where they are going, or they don't hear it enough.

When stakeholders don't get enough support from transformation leaders.

When people don't have ownership of their change.

When energy is focused on something other than the transformation vision and work. Priorities.

When people don't understand the methodology: methodology wars.

Chris' List on Why Agile Fails

Product owners who are not from the business.

Belief in the myth that agile saves time and money.

Launching agile and scrum teams in haste.

Agile roles not embraced or skipped over.

Leaders misunderstanding their role and commitment to agile.

No vision and strategy for people to connect and hold onto when times are tough.

Not training everyone.

I tell Eve it's my narrow view of change, and that I felt very inexperienced making my list. She said that I did it right, then.

My first point is that when agile is forced on people, it sends people in many directions. Mostly, it pushes them away from the change. Eve agreed, adding that agile can't be rolled out or treated as a project. With Agile being a methodology and a mindset, a different approach is needed. One that has a vision of where the organization is going, and supports that with how we're going to get there. One gives people a wide road to try and learn agile within their world, but with guard rails so no one falls away. There is more ownership with this approach, which will ultimately lead to sustained change. I have many questions about this, but Eve encourages me to keep going on my list.

Next, agile fails because there is no single vision for an organization. Lack of an overall vision doesn't give agile a connection to the organization. Without that high-level connection, the outcome is that teams and leaders can't make the connection to their work. Or, it's an inconsistent connection that drives people and systems apart.

Related to this problem is when there is a vision, but there are competing initiatives around the organization. Obviously, this is confusing to people. And it's noisy, too. Many messages and initiatives, so it feels like every group is out for themselves. The outcome is that people break away and begin doing their own thing, missing opportunities for the good of the entire organization. And, they don't listen to any of the messages.

There is one more reason I pull out of my list: people don't understand agile.

"Do you think that means you just get everyone involved trained?" Eve asks.

"I don't think so. I believe it's good for everyone involved to have training of some kind. But not understanding agile can also mean not *wanting* to understand agile."

I have this growing feeling that Thad doesn't want to understand agile. He brushes right over it, leaving it in the shadow of his huge push for lean. If he really is a lean expert, he should understand most of the other methodologies being used, like agile. There is an intent in his actions that I can't place, but it's feeling like a methodology war. The outcome of this situation is that we confuse priorities, people hedge their bets and pick sides. There is probably much more mayhem to this situation, but I don't have any ideas.

"Joel, we've discussed how agile adoption isn't about how well you practice agile, or about culture change. It's about the outcomes. A company has specific business outcomes they want. Whether they are backed in a corner and in trouble, or successfully looking for the next innovation, senior leaders have outcomes in mind. So a vision is built to set the direction and reason for change. Strategies are put in place to begin the change. So when you have a vision, and a strategy, you can look for the outcomes as a measure of the change."

"I suppose you have a fancy name for all that."

Eve smiles slightly, "I guess you can call it fancy. *Hoshin Kanri.* It's considered a lean tool because it's originally from the Toyota Production System. It's a method of deploying strategy from the executive offices out through the rest of the organization."

"I see. So, it's probably something I need to dig into in the future."

"Definitely, Joel. And from an agile perspective, it's a system of feedback loops to make sure the vision is understood, and sensors are in place to test whether or not it's working, and letting people closest to the value stream figure out the best way to achieve it."

"This is great. I think we can really use that," I say. "But can we back up a bit, so I can make sure I understand the big picture?"

"Of course. You know this by now, there is always another frontier to learn. Uncertainty will always be with you. What you do with it is up to you."

I shake my head and smile, "You're killing me."

"Big picture, Joel…" Eve waits.

"The vision guides us, the strategy pushes us to take steps now, and the outcomes are how we measure success."

"Yes," Eve is still considering my statement, "I'm not sure how I feel about strategy *pushing* us, but I think I'm hung up on the word. I like the meaning of what you said. Nicely done, Joel."

Eve observes WL is doing some of this right because we have some kind of foundation in place with the overall corporate vision, *grow happy, grow strong.* She assures me that this is a healthy start, even though it might not feel that way right now. But we could be doing more, and some of the things I described are concerning."

We take a break for a moment.

"So, these are some good observations, Joel. And some interesting things happening at WL. Including the methodology war. It's so typical."

"Really?"

"Yes, many companies struggle with the same. It's a thing; you might want to research it more."

Eve reminds me of the fiefdoms and empires that were built up through the current waterfall and project focused work. New methodologies often disrupt this for the better, but also make room for new empires, so be cautious of that.

Yes, indeed.

The more important point is about combining methodologies to yield a better outcome. Eve believes I already understand this. The methodologies of agile, lean, DevOps, continuous development, and others are nominally different. They are not just methodologies, either. Collectively, they are a mindset. They are not just for software development, but for helping businesses of all kinds be successful. Winning. Imagine when a combination of methodologies and mindsets are used together, they are a force multiplier for success. There is some research on this as well.

"Huh. So, we are a statistic."

"Yes you are. It's okay. If everyone was buying into one methodology, hook, line, and sinker, we would have a different, more concerning problem on our hands."

"I have a lot of reasons for why this transformation may fail. An ongoing methodology war is concerning, even if typical. It seems like we really miss out when we do this."

"Yes."

"I'm also encouraged that our transformation team is learning all the different methodologies that we may use at WL. We can offer a collection of best practices that help solve the problems that emerge. I think that's the ultimate challenge for leading the transformation, to find a way to do this so ultimately it becomes a way we use it everywhere at WL. Our transformation team must be that force multiplier instead of a myopic group that has one answer for every problem."

"That's a great observation. What else?"

"Uh, out of all the things on this list, vision is king. Vision trumps all. Most of the reasons this transformation can fail are likely rooted in vision."

"Tell me more."

"Well, having a vision, communicating it well, and helping people understand how they can connect their work to it."

"What else?"

"Vision…takes many leaders who are in alignment. I have to learn how this transformation can resonate across WL. Working with business partners is good practice for that."

"Good. What else?"

I am exhausted. Again.

"There is more?"

"You tell me."

I reflect, and I rest.

"If I don't focus on vision, and get others to as well, none of the other reasons agile fails matter."

Eve sits back and waits. Silence. She looks satisfied with my answer, but she's not saying anything. We're not finished yet.

I can't stand it. I have to ask what she's waiting for.

Eve leans forward, "If you don't focus on vision, and get others *to care about it* as much as your team does, none of the other reasons agile fails matter."

"Don't just make them buy the vision. Get them to own the vision. Because their work is on the line too," the words are just coming out of me…

"Success isn't in the transformation team, it's in WL. With teams, leaders, facilities…all of it."

"Right," Eve leans back and smiles.

We are both quiet for a moment.

"You're racing soon, Joel," Eve grins. "I'm not giving you an assignment this week."

"Thank you."

"On race day, just by finishing, you will learn something deep about yourself, and perhaps about this transformation."

Reflections

Using methodologies and mindsets together can be a force multiplier for change and innovation.

Vision is the king of change.

Impossible or Unlikely?

They say doing something that scares you is good for you. That sounds good, until it's the day of your first Ironman, or the day I agreed to lead the transformation at WL. Then, the decision to progress forward seems dumb, even insane.

So here I am, facing that moment: race day for my first Ironman. The parallels between my work at WL and my work for this race are incredibly obvious: a huge goal, lots of risk, it takes forever, people don't understand you, and coaching is supposed to help. Of course the risk of the WL transformation is much greater than my Ironman race. Yet preparing for today took so much time and effort. The majority of the effort was in actually finding the time to get ready for the race. How to fit in the training? I had to create the environment in my life to make room for the training, and the recovery from it.

That effort seems similar to transformation work. It's also very time consuming, and the majority of my effort in preparing for the change. The steps toward change have been a challenge, but not nearly as difficult as creating the environment for it to thrive. People and teams will figure out how to transform their work with lean, agile, and DevOps practices; we need to make sure they have the environment to be successful.

Walking into the transition area of the race, I'm relieved. Still nervous, but I feel like a big weight is off me. The pressure leading up to day is gone, because finally, it's race day. I'm no longer worried about the training I missed or didn't do, I'm no longer wondering how I'll get my next workout in. My sole focus is to enjoy today through three sports, and eat. And to kick The Mouth's ass. I got this.

Race day was a typical, foggy August morning. In fact, it was so foggy that the swim now looked absolutely daunting. I'm not that great of a swimmer, but I've trained well to be able to go the distance without it taking too much out of my energy stores. But standing on the shore, seeing the swim buoys disappear in the fog, I was secretly hoping maybe the race organizers would cancel the swim. How will they see if anyone is struggling? I'm going into a slight spin on this problem. I have to think positive thoughts. The fog will burn off. It always does. It will lighten up by the time we swim.

The race begins with a time trial or rolling start, so not everyone starts at the same time. Swimmers are supposed to seed themselves based on their swim finish time. Faster swimmers to the front of the start line, so they don't have to swim over everyone. Fine by me.

While finding my place in line, Tony comes up to me. I get the same old punch in the arm, before he looks at the finish time for the group I'm in.

"You ready Joel?"

"Yes. Are you?" Anything to get this guy away from me… Wait, that's not the right mindset. I can do better than that. I'm going to kick this guy's ass today; I should get a good look at him. I should enjoy this moment instead of wishing it away. This is the magic of positive thinking working in my favor.

The fog did lighten up and visibility was fine. Once I got swimming, though, I wasn't fine. The longer I swam, the worse I felt. I don't know what happened to me. I ate well the days leading up to the race, and I had a good breakfast this morning. I'm physically ready. Mentally, I'm ready to stamp my victory over The Mouth.

Whatever the reason, my swim was a real challenge. That's when negativity crept into my head, and had a big party. *What's happening to me? Maybe this day was a bad idea! Maybe I should quit now, so I don't drown! If I can't do this short part of the race, how will I ever ride 112 miles? And a marathon after that! This is not working…*

On it went. But I kept going. I thought of my family and of Eve, and didn't want to tell them that I quit because I didn't feel good. Adding to the turmoil was the seaweed problem. The north end of the lake is very shallow, and the area is thick with seaweed. Every arm stroke is full of weeds. I'm not squeamish with lakes, but swimming in this thick, green sea forest was disgusting. It was stuck to my goggles, and who knows where else. On top of that, I couldn't help but think that I might get tangled up in this stuff, and not get out of the damn water. The mantra I had planned to use, *my race*, was long out the window. This was about survival.

Two hours later, I made it. I looked like a swamp monster; I'm sure it will make a great race photo. I was disoriented, and had to walk to where my bike was in the transition area. My head was pounding with pain. I wasn't sure I could even get on my bike. My thoughts were irrational and mixed up, and I knew it. Time to pull it together. My coach prepared me for moments like this; I just didn't think I would have one so early in the day.

I drew from our conversations of breaking this race down into small, manageable pieces. Don't look at the whole race, or it will be overwhelming. And, when things don't go as planned, thinking in small chunks makes it easier to let go of the bad moments. So I had a bad swim. It was time to let go of that and get on my bike. It's just a bike ride on a beautiful, summer day. And, I have to beat The Mouth.

Speaking of summer days, today Colorado served up a hot one, with a high of 98F. At the beginning of the bike, of course, it wasn't that bad. But it didn't take long for the bike to get windy, and hot as hell. I'm no longer thinking I'm going to beat Tony. In fact, I'm being passed by people on the bike. I didn't think today was going to be me passing people on the bike all day long, but I didn't think I'd get swallowed up by what appeared to be nearly everyone out on the course. How could there be anyone behind me after my two-hour swim? It was a long, dark moment. I really did take forever in the water… *Stop. Let go of what happened in the water. Ancient history.*

So now I'm early in my 112-mile bike ride, and I don't even know if I can make it. I'm dying. Why am I doing an IM at altitude? Why am I racing when it's 98F? Why is this challenge so alluring?

I have this bike course in parts. I need to remember that. This is one part of the bike course. It's unpleasant and killing me, but I will survive. There is a light at the end of this tunnel. I just need to make it through highway 36…

I'm still getting passed. Finally, the biggest hill is over, and we got an aid station with food and water. What's the next small part of this race? Getting to the flat part. We are almost there.

Close to the flat part of the course, things changed a little. I was actually passing and getting passed by the same person. The athlete would pass me on the uphill, and I would catch them on the downhill. Repeat. This was my first real joy of the day. Slowly, I was able to come out of that dark place.

Then, we hit the flat road. It's a good 25-mile section going east, but of course today it wasn't just hot; we had a massive headwind. But for whatever reason, my funk was lifting. I'm getting into a groove, even with the wind so relentlessly strong. This is where I began to really feel better, and began passing people. Yes, I'm getting in the groove…

I arrive at a bike course turn around, and my family is there to cheer me on. I didn't think I would see them in the crowd, but there they were, going nuts as I rode past. It really motivated me to keep rolling.

One moment I will never forget was as we approached Longmont, some kids were handing out freezie pops. They were grape, and they were life changing. There was a sticky, slippery mess on the road in that section, but it was totally worth it. I was so dehydrated from my effort, and the hot, windy conditions, the freezie pop seemed to instantly evaporate as it slid down my throat.

I was still feeling good, but the conditions of this day were starting to wear me down. I couldn't wait to get to the halfway point and access my special needs bag. This brilliant idea allows athletes to put whatever we want in the bag: food, sunscreen, you name it. It's a great contingency option for athletes to plan for success, on a day when many things can go right or wrong, because of bad decisions or just rotten weather. Today, we most definitely have the rotten weather conditions. Most athletes pack extra food or drinks in the bags. Having a great day and don't need to stop? No problem, don't stop.

I had some fun things in my special needs bag. When I packed it last night, I wasn't sure I would use it, but after a few hours of hills, wind and heat, I most definitely needed more calories.

Rolling through the special needs area, I didn't know what to expect. I'm a newbie to all of this. I was shocked at how well the teams of volunteers worked together to get each athlete's bag. There were not a ton of volunteers, but there were enough to help each athlete that wanted to stop and get their bag. There was a spotter up the road calling out athlete numbers, which triggered the team serving that number range to jump into action, grab the bag, and get out to the road and hand it to the athlete. The athlete can stop to get the bag, or ride through. A process of small batches, and few handoffs.

Many athletes grabbed their back and kept riding; I planned to grab a bunch of ice and dump it in my jersey as I rode. But conditions made me adjust this plan. Once I got my bag, I stopped. Going a little slower now might help save me in this heat. I poured an entire bottle of water over me, and then packed a bunch of ice in my bike jersey. It's a heart stopping experience.

Then I unwrap my prize: a grape soda. As I did, and poured it into my water bottle, another athlete watched. Also, I had a homemade burrito and a bag of potato chips. I stuffed the burrito in my bike jersey pocket, and held the bag of chips.

"That's the best idea ever," he stared at the can, and then at the burrito, and then at the potato chips.

I took a sip. Heaven. I confirmed to the athlete that he was right, wished him a great day, and got rolling again. A mouthful of potato chips, I'm packed with ice, and I've got a burrito in my jersey pocket and a grape soda in my water bottle.

The calories were improving my mood even more, and my pace was now strong. Even though I'm eating and riding, I am passing people like crazy. Yes, I am moving up the food chain of this Ironman, and loving life.

At this point, we had a cross wind, so we had a reprieve from the draining work of riding into a headwind. The tradeoff was that this part of the course has no shade. The heat was no longer just a hardship; with us athletes getting baked in it all day, it was a real problem. I begin to see athletes stopped on the side of the road, obviously blown away by the day's effort in the heat. And there were several crashes in these later miles. Just imagining that people were losing their focus and then unable to ride safe.

I pressed on, and entered another dark time. Though I didn't feel nearly as bad as I did during the swim. This was different. My calories were wearing off. I had more food with me, but didn't think to eat it. The loopiness I felt from the heat, wind, and altitude stole my ability to focus. I was aware that I was losing my edge, but couldn't do much about it.

If I thought things were difficult before, I was naïve about my race. The crashes are haunting me. There are so many turns on this course. Couldn't they get better roads closed for this event? We had this huge climb on highway 52, and then suddenly a 90 degree turn, and then we had to climb a bunch more hills. One of the hills was pretty steep; so many people fell over because they couldn't get themselves up the hill. I'm just a regular athlete, but I did have a technique from my coach called S-turning, that helped me get up the hill without falling.

Then I hit the aid station at mile 83, and saw Cele and the kids. They gave me a mental boost that I desperately needed. I entered a more hilly section of the course, and now the wind seemed to pick up again. Nature just wanted to thrash us today. As I'm pushing through this difficult time, a woman crashed just 25 meters ahead of me. I don't know what happened, but she went off the road onto the shoulder and then went flying off her bike. A bunch of us athletes saw it happen, and two of them stopped to see if she was all right. I stopped too, but they waved me on. They knew her, and wanted to help.

It's so very difficult at this point. I want to quit. I want to sell my bike. I never want to do another Ironman. My own negativity surprises me.

Another bright spot opens: I see the other athletes using my coach. We don't train together much, because we're all busy with work and our families. Today is our day, and lucky for us, we found each other late on this daunting ride. We can't legally ride together, but we catch each other every two miles or so, and offer encouragement.

Cranking up one of the hills, there are a ton of spectators lined up on the hill. Just like the Tour de France. Well, similar. I see them and the hill and I'm just not sure how I'm even still pedaling my bike at this point. I make my way up the hill, and the hundreds of people there give me mojo and mental strength. Then I see one fan dressed like the grim reaper, drinking beer. I laugh aloud. For the first time in this day, I am having a good time. How ironic that it's happening when I'm least confident in anything I'm doing.

With less than ten miles to go on this 112-mile bike odyssey, I'm very aware that this effort is now literally taking everything I have. My only goal was to keep going and not give up.

We're in the absolute heat of the day, there are no more aid stations, and I'm just about out of water. Approaching the airport, I know there is one more hill to climb. Stupid hill. Yes, that's my current attitude. It's at this moment that I see The Mouth. We yell at each other, and I'm sure if we were any closer together, Tony probably would have reached over to punch me in the arm. If he did, I would have fallen over. At least I avoided that today.

Seeing the look of dismay on his face that I was right there with him was priceless. It was the ultimate high of the day, and I will never forget it as long as I live. In all the struggles on the bike course, I forgot about kicking Tony's ass. Huh.

Shortly after that, we are dismounting our bikes together and heading to our bike-to-run transition. As I park my bike, I'm certain I don't want to ride it again. Ever. Yet I am right there with The Mouth, a seasoned Ironman. My hate for my bike is quickly overshadowed by the fact that I caught The Mouth.

In the changing tent, getting ready to run a marathon, I'm overwhelmed. Again. I don't think I can do this. I know I just said I'm going to kick The Mouth's ass, but I am so demolished, and it is so hot. I'm out of my mind with dehydration and exhaustion. With a bunch of other wasted athletes, I sat there for several minutes, contemplating how to get up. Somehow, I find a way to stand up, and start walking out of the tent. I see just a few athletes run past, on their way to the finish. Wow, people are finishing, and I haven't begun my marathon. I don't know where The Mouth is, and I don't care.

The two-loop course begins with a downhill. It sounds like a good thing, but it has me worried. How will I ever make it back up this *same hill* in 13.1 miles? And, again at the finish in 26.2 miles? It seemed just about impossible. That's when The Mouth trots past me. Whatever. I've got to survive this my way.

Survive it. Ugh. Yet another dark moment is upon me. Why am I doing this, anyway? I actually paid big money to be out here. All that training, all the gear and the logistics…and here I am, barely able to make forward progress. The mantra *my race* means nothing at this moment. I'm not worried about others. Except maybe I'm concerned about someone flopping over in front of me, and being too exhausted to be able to move out of the way in time.

Why am I here? What's my one thing? My one reason? It's not to beat The Mouth. That may happen or not, but that's not why I signed up for this mayhem. I registered for this because I wanted the challenge. I am here to challenge my limits. That's my one thing. When I challenge my limits, I can explore new places within myself. What am I made of mentally? What can I do physically?

So if I stop or quit, it better be because I'm at my absolute limit, and cannot take even one step more. Am I at my limit right now? No. I can keep going. So I will.

The shade on the Boulder Path doesn't last long. I'm baking in the sun again. And, this is Boulder so all manner of human civilization is out *en masse*. Kids are zipping past me on skateboards and bikes. There are hippies and several coaches are riding on bikes next to their athletes. Mine was somewhere in the mix racing. There was even a makeshift (unofficial) aid station that offered kinds of liquor and pot. The man running this aid station was definitely crazy. The wildest things were happening, and I was experiencing it all in slow motion.

"Pretty chaotic out here, isn't it?" A cyclist says as she pulls up next to me.

At first, I don't look at her. She must mistake me for someone else. After all, we're all in sunglasses, caps, and skin-tight outfits. This could be awkward, but I'm so hot, tired, and out of it, I don't really care. Finally, I turn to look at the cyclist. Holy crap, it's Eve.

"Eve!" I am breathless.

"Did you stop and get some pot?" She's grinning. "I hear they have some really good stuff."

"I—I'm shocked to see you out here."

"I may not be your triathlon coach, but I can still kick your ass out here."

I sort of laugh, "It's great to see you. Just don't make me rearrange the process at any of the aid stations."

"Well, everything can be improved, Joel. So they probably could use your help. Speaking of improvement, how are you doing? Have you done a mental assessment lately?"

I hadn't. My coach trained me to do that, but apparently, not very well.

"I take that as a no. So, let's do one. Scan your body from top to bottom. Anything hurt or feel tight?"

I do as she asks. Everything feels okay. Not great, but not painful. Well, actually my ankle is stinging, burning. That can't be right. It's the ankle that has my timing chip on it.

"I think I'm okay, except my left ankle. I think my timing chip strap must be too tight or something."

Eve tells me it might be a good idea to stop and fix it.

"Brilliant," I say. Wow, I'm so out of it. I pull off to the side of the path, and look at my timing chip.

"Looks like your compression sleeve for your calf is over your chip. That's not good, Joel."

I make a calculated move to bend down and adjust my calf sleeve, so it's above my timing chip strap. I also open my chip strap to loosen it, and then close it around my ankle again. My hands are sweaty and slippery, and my legs are shaky, so it takes a moment to get it right. I don't care because it feels good.

I stand up and begin shuffling again. There is instant relief.

"I feel so much better now."

"Great."

Eve is so upbeat and positive. Far more than our usual coaching meetings. I am a happy recipient of any and all of it.

"Now, do that mental assessment again. I'll be right here while you do it."

I do it again. "I think I'm good, now."

"Okay, that's the physical part. How about your head? How's your attitude?"

"It's in the toilet."

Eve laughs, "It can't be, or you wouldn't be running. This is a killer day, Joel. People are giving up all over the place. You're doing very well. So tell me, what were you thinking right before I pulled up next to you?"

"Okay, so my attitude isn't terrible."

"What were you thinking?" Eve persists.

"I was thinking about how The Mouth just passed me. I want to beat him, but at the moment, I can't answer his pace. I have to focus on my own effort." I start to get fired up again.

"The Mouth," Eve laughs again. "I can't wait to hear about this competitor. He has climbed all the way in your head, Joel."

"He is something else. Maybe Thad's lost twin."

"So what else were you thinking?"

"I was thinking about how impossible it will be in 13.1 miles to run up the hill I was just running down. Well, shuffle, jog, whatever. And then, to do it again at the finish."

"Impossible, unlikely, or just really hard?"

"Huh?"

"Were you really thinking that running up that hill was going to be *impossible*? Or were you thinking it was unlikely? Or, were you just thinking it was going to be really hard?"

I laugh, a little breathless. After all, I'm doing a marathon at altitude.

"Well?"

"Great point, as usual, Eve." I shake my head. This woman never ceases to find my mental tipping point, and then shove me over the edge into the place I'm supposed to be.

"The people you will see walking or even lying on the side of the bike path today are those who believe finishing today is impossible. Maybe medically, for some of them, it is. But for many, they just can't fill the gap, or make the leap, in their heads. It's the same for your agile transformation as it is for this marathon, Joel. It seems impossible for you or WL to change, but it's really just unlikely."

"Unlikely," I repeat. "Yeah, there is a lot of today that feels like that, too. What's your point?"

"It's unlikely for everyone here, Joel, for different reasons. That's part of the mystique of Ironman for most of us, if not all of us. We are nervous, maybe even freaked out on race day morning, no matter the conditions. Because you might not finish. All of your training, all of your preparation, all of your gear…might not help you get to the finish line. And yet, you are here. Think about it, Joel."

Eve rides away, calling over her shoulder, "I'll expect an answer later today."

I'm left chuckling to myself. That's my coach: making me work on my mindset, even during my Ironman. Why would I ever expect anything less today? I should have asked her how long I had to think about it. Will she show up in one mile, or 12 miles? I can just picture her camped out before the big hill, requesting my answer. I can also picture her hanging out *after* the big hill, asking the same question.

Impossible, unlikely, difficult. What the hell. I know the answer, but I gotta think about this.

After a few miles of shuffling, I try to eat more and get into some kind of pace. Then I start to see the real carnage of the day. Eve was right. Strewn along the side of the path are all kinds of athletes who had just given up. They were lying in the shade on their back, or puking from dehydration or altitude sickness, or both. In a weird way, this got me going. *Unlikely* got me going.

I wasn't going to come this far and quit. I also wasn't going to make it this far, and end up in the hospital. I was going to finish today, and be in my own bed tonight. Be smart, pace myself, and never give up.

I'm back to breaking things down into small chunks. Originally, I thought I would take the run a mile at a time. Not today. I focus only on the next orange cone marking the course: run to that cone, then walk to the next one. Repeat. The more I can run this thing, the faster I will be done.

I do another scan of my body and mind. Does anything hurt? Am I drinking enough? Am I at my limit? Is there anything else I can do right now? Would I regret tomorrow the effort I'm expending today? I determine that I'm at my absolute limit. I can't do anything faster or smarter than I am capable right now. I won't have any regrets about my effort right now, either.

I have my answer for Eve. It's unlikely I'll run up the hill, because it's at mile 13.1 of an Ironman marathon on a 98F degree day at altitude, it's my first one, and it could take all of my energy and I'll have none left to finish. There. Hopefully, I will see her before I forget all of this.

It looks like it could storm. A typical August afternoon in Boulder, but please not today. If a storm whipped up, the run could be canceled because of safety concerns. My body wants to stop, but my brain very much wants to have this accomplishment today. I'm too far in to have it stolen. I want to be an Ironman.

Working hard at running, I hear this strange noise. I thought it was the guy next to me, but when he stopped running and started to walk, I could still hear the noise. Was it my shoes? I must be losing it. Moments later, I realize the noise is the ice cubes bouncing around in my jersey as I run. Huh. I've lost it, for sure. Hopefully, I have enough mental power to make it to the finish line.

This is when I see The Mouth, heading the other way and looking great. He's practically bounding compared to everyone around him, including me. Crap. He has to be two miles ahead of me. I don't think he saw me. I would have heard his big mouth say something annoying if he did.

Seeing Tony flipped a switch in my foggy head, and turned me into a competitor. It remained with me the rest of the run. I felt miserable and great, all at once. This new flame inside me pushed me past what I thought was my limit. The reality was that I wasn't at my limit yet. Nothing like an annoying acquaintance to snap things into perspective.

I begin peeling off the miles, and I'm actually running most of the time, and walking through the aid stations. The longer I'm on the course, the more medical situations I see. At one point, I see 10-15 guys who look just like me, sitting there on the side, absolutely done. Some of them have emotion left on their face, like dejection and anger. Some just look like exhaustion is the only thing on their mind. Many people would argue that lasting this long in an Ironman on a day like today is not a failure, and is in fact impressive.

Medical support is scooping people off the road as fast as they can. Several times, I overhear conversations between athletes and family or friends. Most are about where to pick up the athlete, who is quitting right there, or who will be at the medical tent.

Meanwhile, I have 15 miles left of my marathon. I'm taking stock in this moment, and I'm grateful; I'm actually feeling pretty good. My run continues to get stronger. Am I over confident, overdoing it, or maybe delirious? No, I think I'm okay. I begin calculating the time it will take to finish this race. I might be able to finish before 10pm. What an introduction to my first marathon and Ironman.

Around six or seven in the evening, the oppressive heat is releasing its grip on us. There are still athletes who had given up everywhere. And now we have a new obstacle: there are bugs and mosquitoes attacking us. They are in my mouth, along with insect repellant that seems to be hanging in the air.

Cele and the kids find me for the first time, and offer me much needed encouragement. They do this between swats at the mosquitoes. They are concerned about me; I tell them I'm fine, that I have no plans to be hauled away by medical.

I'm approaching the turnaround point, mile 13.1. I walk-run up the giant hill that I ran down at the start. How will I run up this hill a second time, 13.1 miles from now? It's getting dark; by the time I am back here, there won't be any daylight. Great. Exhausted and running in the dark, up a giant hill. I can't even imagine it. There are families all over in this area, and I overhear some people talking about the temperature was 98F today.

I see Cele and the kids again. They are so very excited for me, but I can see concern on their faces. I'm sure I must look how I feel. They have special friends with them: Meryll and her kids. They high five me, and I have a fresh burst of energy. I'm so jazzed to see them. Even though I chose to do this race, it has become a lonely proposition. I didn't realize just how lonely I was until I saw this group.

Meryll is going to do this race in the future, I just know it. I can't wait to talk with her and Jack about it later. Who knows, maybe Jack will get the bug, too.

There is a well-organized special needs area at this turnaround point on the run. These volunteers are amazing. Their process is efficient, and they are not acting like they've been working in 98F degree heat. I get my bag, and begin to enjoy another grape soda. I walk and sort through my goodies: cheese curls, and another homemade bean burrito (without cheese). No need to heat that thing.

Then, I find the best part of my special needs bag: my Mike & Ikes, Skittles and licorice. I had forgotten all about them. I was never so happy to see these little bags. Yesterday morning, when I packed this bag, Caroline insisted I separate them by flavors. She was convinced that some flavors would taste better than others, and I wouldn't know what would taste good until I got to this very point in time. "It's all about options, Dad."

Was it ever. Orange looked and sounded disgusting, but lemon and lime were revolutionary, ground breaking flavors. I kept the cherry bag and licorice for later. Options, sweet options.

"Nice job on the hill, Joel."

"Thanks." I turn to see Eve on her mountain bike, riding next to me. "Kind of dark to be biking, don't you think?"

"I'm fine. After all, I have better lighting than all of you."

"Yeah, we should have headlamps like yours. It's going to be very dark very soon."

"So, do you have an answer?"

"Unlikely. Not impossible."

"Tell me more."

"Like you said, finishing this race is unlikely any day of the year. Today's conditions make it even more unlikely that I will finish. That's why a lot of people don't do Ironmans. They don't want to go through with all the time, effort, and expense without a guarantee that they will finish. Just the swim is really hard (well, it was for me). Just the bike ride is very hard. Just the marathon is very hard. As single events, those are really popular. They're challenging. But they are not an Ironman."

"What else?"

"When I break this marathon down into chunks, it doesn't feel impossible, or even unlikely. Just…hard. It's hard to run all the way to the next aid station, but I can do it. It's hard to run up that damn hill, but I can do it. If I can focus on these very hard small pieces, I will conquer the unlikely. I will conquer Ironman Boulder."

"Excellent. Have you done a mental scan lately?"

"Yes."

"Good. You've really picked up the pace. Most people are slowing down or quitting. You're really nailing this day."

"Thanks."

I told Eve about the moment when *unlikely* clicked for me, and actually became a motivator. I want to be one of the few people of the world who have completed an Ironman. I want to be one of the rare people who can finish Ironman Boulder *today*. And, I want to be one of the rare people racing Ironman Boulder today who can actually run the marathon. I chose this goal because it was exceptional, and now I'm running my way to it.

"The sooner you finish, the sooner you will be one of those unlikely people." Eve knows just what to say.

"So I will."

Eve pedals away from me, and I'm left working on my next small chunk of difficult.

Well into my second loop, I see The Mouth going the other direction. He is walking, and he looks terrible. I can't believe it. Mr. Big Shot, who was bounding past me at the start of this marathon, is walking. I think I am ahead of him by at least four miles. For just a moment, I wonder how it happened. Then I no longer cared. The raw reality of the moment only demanded I think about moving myself forward.

"Joel!" Tony croaks at me. "You're on your second loop, you bastard!" He would punch my shoulder if he could reach me.

"Let's do this, Tony."

"You're at least four miles ahead of me! It's your day, my man."

"We got this," is all I can offer.

Well, that got me going. I have a long ways to run yet, but it looks like I'm going to beat The Mouth. Now I have a new goal: I want to beat him by a clean hour. Yes, I want to decidedly, 100% whoop his ass.

It's now completely dark. This part of the marathon course has no lights, so I can't see a thing. I heard people were injuring themselves, tripping on the path because it was so dark, and because they were exhausted. I am wearing a glow stick, but that only helps people see me. I can think about being careful, but I can't really follow through on it. All I can do is will it to be true. After all the hardships of today, I can't accept failing because I wiped out in the dark.

I have three miles to go. I haven't seen Tony, or any of the other athletes I know. Then again, with it being pitch black, how would I know? It's sprinkling, just enough to be noticed. Heat, wind, hills, bugs, rain, darkness, and exhaustion. It's all stuff in my way, and I'm doing it anyway. All I can do is focus on forward progress. My latest scan reveals only one thing: I'm so out of it, I no longer feel hot or cold, only tired.

Three miles. I check my watch. Actually, I have to study it with intensity, because my mental faculties are so slow. If I hustle, I think I can make it in under 14 hours. The length of time isn't what I'm focused on. In fact, I don't even comprehend that I have been moving for almost 14 hours. I'm focused on how close I am to the next hour. Now I have a new goal. Once again, I have found the energy I need to push just a little more, just a little. Refreshed, I think of Eve. She would love how I adjusted and found a new goal to persevere.

I'm nearly there. I am going to make it. I can't believe it. At this point, I could crawl the rest of the way, and still finish. But I won't need to crawl, I hope.

"So, where's The Mouth?" Eve's voice is next to me in the dark. She's no longer cycling, but running next to me.

"I'm about four miles ahead of him." I can't believe I'm even saying that. It sounds so good.

"Well, things changed, didn't they?" Eve's positive voice is giving me more energy by the moment.

"Must be those nasty Mike & Ikes you were eating before."

"Still eating them." I am breathless and exhausted.

"So now what? You're almost finished, but...are you?"

"New goal is to beat The Mouth by at least an hour."

Eve laughs, "Wow, he must really be annoying."

"Yep."

"How will that happen, Joel?"

"One hard moment at a time. Small chunks."

"So, you just have to keep your pace, and you'll reach your goal?"

"Yes, including running up that damn hill for the second time."

"Excellent. Then you can get him out of your head."

"I hope so."

"Are you holding a bag of cheese curls? It's so dark, I can't make that out."

"Yep. Best marathon snack ever. You should try it for your next one."

Eve laughs, stating this is no time to discuss anyone's next Ironman. It's all focused on me and my big push to the finish. My big, spectacular finish to this ridiculous day.

"Back to The Mouth... You will get him out of your head when you beat him by over an hour. At that moment, he is no longer your competition. You will have a new goal. But we can talk about that another day."

"Yeah, I'm sort of involved in doing the unlikely right now."

Eve laughs, "Indeed you are, Joel. And, what an impressive effort on a ridiculously difficult day. Congratulations."

I'm wary of being congratulated pre-emptively, but it does sound nice. "Thanks, Eve. And thanks for being out here."

"My pleasure. All right, Joel. I'm going to drop off. This is all you, now. Enjoy your finish. You earned every last second of it on this very challenging day. It's going to be fantastic, because it was unlikely."

I thank Eve, and have a renewed spring in my shuffle. At least it feels that way. I have no idea how fast I'm really running. I can look at my watch, but I can't make sense of the data. I'm in this moment, where time matters but doesn't seem to pass. I'm moving forward, but I'm in in one place. It's an incredibly present moment, which might feel good if I wasn't in this exhausted state of suffering.

That giant hill is now in front of me for the second time. I open another mini bag of cheese curls, and realize I can stop eating, and I don't need to carry any of the other food I've got. Remnants of my special needs bag. I laugh to myself about this scene. Running in the dark, uphill, eating junk food, because I want to be an Ironman. Because I *will* be an Ironman. I will find my limit.

I felt like I had been running forever, but then I see some athletes just beginning their second loop. Things could be worse, I guess. Here I am, scrapping to get in under 14 hours, and these people will be at least two hours behind that. Most of them are walking and wearing glow sticks. One man walking congratulated me on my finish, I reminded him I wasn't there yet. He said he had full confidence that I would make it the last 500 meters.

I offered him my fresh bag of cheese curls and the rest of my Skittles. He gladly accepted, and said something about how this food was going to make walking this final loop the best one of his life. Conditions changed so much from when all of us began this race to the current moment. The kind of moment when your leftover food becomes another person's treasure. I scurry along, because I'm very close to the hour changing over. I have worked very hard the last three miles to get under 14 hours; I can't miss the goal now.

I check my watch for the tenth time in a few minutes. It's going to be really close, but I think I'm going to meet my new goal. And the ultimate goal: finish Ironman Boulder. I can hear the announcer calling out people's names and saying *You are an Ironman!* Man, I am really gonna do it.

Everything I've struggled through today fades to the background. None of it matters anymore. All the aches and pains I had are gone, and somehow my legs turn over faster and faster toward the finish. I did it. I really did it. I'm in this place of accomplishment and achievement that feels incredible. Doing this race was the best idea ever. I can't believe I haven't done one before now...

Then I see the glow of lights. Finish line lights. I have goose bumps. *Goosies*, as my daughter Cici likes to call them. This moment is electric; it's the impossible thought meeting reality. It's me at my limit. Yes! I will make it. My heart races faster, and I feel very alert. I am completely focused on the finish line. An Ironman finish line.

I hear my name called out in the darkness... Holy crap, that's *me!* The finish line is right there... The clock is at 13:59:31. Ugh, I have 29 seconds to do this... I got this...

I make it: 13:59:49. Wow.

At the finish line, I can finally stop running. It feels good and bad, all at once. Just like most other things I experienced this day. I can barely stand up, but I'm so jazzed about finishing, that I have newfound strength to go through the finisher corral. I get my finisher medal and shirt, and then pour two waters over my head. What an incredible feeling. I imagined steam rising from my head. It's much cooler now, but I don't think my core body temperature has cooled at all.

My family and Meryll are there to greet me. They offer congratulatory hugs. I'm barely able to hold my balance while they smother me with joy and excitement. Then Cele studies me, asking if I think I need medical help. No, I assure her I'm fine.

"Dad, you did it!" Cici jumps up and down. "And, you have cheese curl crumbs on you."

"Way, to go, Dad," Elliot high fives me. "I can't believe you did it."

Caroline looks at me with her nose scrunched, "Dad, you're a mess."

"Of course. After 14 hours of doing something unlikely, I am a sweaty mess."

"Thirteen hours and 59 minutes, Dad," Caroline corrects, and then looks at me for a moment. She's probably trying to decide if I'm out of my mind or not. She's not alone; I'm still trying to decide that as well.

"Unlikely." Caroline smiles. "That's a good word for this insane event, and for doing it on a day like today."

"The only one that fits."

Reflections

I found new, good parts of myself. I found new, extraordinary limits.

Sense and adapt is king for an agile mindset and Ironman races.

I'm more competitive than I ever imagined, thanks to The Mouth.

Setting small goals inside a huge one gives short term purpose and motivation to reach the overall goal.

Unlikely: how does this fit with WL's transformation?

Culture Shock

Two days after Ironman Boulder, I'm back at work. With our cyber security work in full swing, and our transformation team getting off the ground, I can't stay away.

Ahead of my Ironman, my coach and I created a strategy to help my legs and body begin to recover that first week or so after the event. I was to do things like walking or swimming, even riding, as active recovery to increase blood flow to my muscles. Increased blood flow helps speed healing and reduce soreness. It seemed like a great plan.

Fast forward to now, and you can throw that plan out the window. Yesterday, I could barely do much walking, let alone swimming. I was tired in a way different than I expected. I had this deep feeling of systemic exhaustion. My entire body is drained from the effort of doing the unlikely, in the heat, wind and altitude. I taxed my whole system, in ways I have never done before. And I'm probably still a little dehydrated.

Going to work probably isn't going to help me pull out of this, but resting when I can, and drinking water, instead of J&L's medium roast, will. I have to take care of this exhaustion if I ever want to get to the increased blood flow strategy.

I'm seeing a parallel with WL. We can work on symptomatic things, like falling behind on product development, or lack of business partner alignment, but the changes won't stick. In fact, they might just collapse under the pressure of the culture or the legacy systems and technical debt. Only when we work on the systemic part of WL, the environment, will our changes stick. The big system must be healthy enough to work on the smaller things. Huh. Who knew Ironman recovery and organizational health had similarities?

I'm getting ready for a phone call with Lucy from facilities, when a new email pops into view, from our corporate communications area:

Culture Corner | First Edition

A newsletter designed to help us change the culture at WL, one person at a time…

Welcome to the first of many editions of Culture Corner. We've got a great experience on tap for you, if you're willing to change! Speaking of that, we're working on changing the culture at WL so we can *grow happy, grow strong.* It starts with each one of us…

Since we haven't had a second C² meeting yet, someone must have gone off on their own. I see a few names from the sales area copied on the email… It's time for my call with Lucy, so I have to stop reading.

"Hi Joel. I'm sorry I had to cancel our last call. I've been so busy, I couldn't keep our meeting."

I assure Lucy no harm is done, but I did want to know if she talked with Karen about joining our transformation team. She said she did, and was excited to be thinking about being a part of changing WL.

"But first, I have to read this C² newsletter. Oh my, Joel. Just tell me one thing…"

"Yes?" I like Lucy. A breath of fresh air outside my corporate box.

"You're on the C² committee. But did you know about the newsletter?"

"No. Although I'm sure it's got good intentions."

"Right. But I'm scrolling through it and see no intent. Oh well."

"So, back to our time together, what's up?"

"I meant to respond to you, but time sort of got away from me. And then my team had a discussion about the scrum teams and how they are writing on the windows. I lost all focus after that."

"Really?"

"Yes. I investigated. They also have giant sticky notes and other things taped on the windows. They are very organized, but it does look like chaos. It's harmless; they were writing with dry erase markers, and using painter's tape. But one of my team members had a fit in our weekly meeting. You know about our rules. And apparently a few of my team members have been walking around by the teams' space after hours. It's like they are on a witch hunt, searching for the bad."

"Sorry about that. I can—"

"They had a fit at *me*, because I was supposed to *reel in this team* and *set them straight*. I have to be gentle with my team, but they were, and are, being completely unreasonable. I knew if we talked about it any further, our conversation would be an argument and I would hurt your cause. So I agreed to take the topic forward, to you and to my leaders. This satisfied them."

"Is the window incident the only problem?" I'm afraid to ask.

"No," Lucy sighs. "I've had two phone calls from the managers of teams who sit near the scrum teams. They say that the scrum teams make too much noise. They are always talking and walking around. Lots of movement and voices."

I ask Lucy what kind of work the complaining teams have. She tells me they are consultants, which is nothing remarkable, but they do spend a lot of work time consulting on the phone with our sales people. They often lead calls and study groups. They have rooms to conduct the calls, but sometimes the rooms are full, and they have to do the calls at their desk.

The scrum teams note that because these teams are always in the conference rooms, they are forced to keep their work on the windows and walls. It sounds like it has become a vicious corporate circle.

"Why on earth were these highly collaborative teams put next to teams that are on the phone?"

"We didn't have space anywhere else. Well, we had some space up on the fifth floor, but the other half is a director team, and they didn't want the noise by them. So these sales support teams drew the short straw."

"Actually, it sounds like the scrum teams did."

Lucy laughs, "That's true. At any rate, I've been helping my team with understanding what these teams are doing. I've been mildly successful; three of them are going to Essence of Scrum training."

"That's a great first step. That should help them understand a little better than now." I resist just saying it's great news. I'm working to focus on the value of the training.

Lucy tells me she talked with the scrum masters again, and suggested that they try to keep it down. It didn't go over very well with the scrum masters. They were firm on their stance that they made it very clear that they would be a collaborative team, talking all day long. She asked them to understand that the sales team had no say in the scrum teams being located by them.

"The scrum masters put in a request to be moved."

"Good for them. How long will it take?"

"I don't know, Joel. It's not that simple. We don't have a big appetite to reconfigure workspaces. We have to get all different furniture, and it's expensive… You know all the talk about strategic expense management. Facilities always gets nailed first because everyone can see the money being spent."

"Good point, Lucy. I never thought of it that way from an expense standpoint. Project work has been mostly out of sight," I sigh. "So, do you know *why* they were writing on the windows?"

"They said they had nowhere else to show their work. They have just one wall for their giant sticky notes. They were trying to map out some things that needed big space. The whiteboards were small, and they were in the conference rooms being used by the teams leading the sales call."

"This is why we need you on our team, Lucy. We can help each other, and then your team members won't have fits and stress you out. And, we can use some Six Sigma coaching."

Lucy agreed. I would have Marilyn add her to our work meetings.

"There's more," Lucy smiles.

"I'm afraid to ask you, Lucy."

"So…we have these highly collaborative teams. They talk a lot, they move around, and they need space to create and discover their work. My team is hardly at their desks, and is also very curious about this new way of working. And, I have this work beginning with you and your transformation, Joel. It seems like a perfect opportunity for an experiment."

Lucy wants to build experimental space for the three scrum teams next to the facilities work area. They have open cubes near them that no other teams have wanted, because the facilities teams are situated on the first floor of the building. The space has always been undesirable, because it is halfway below ground level. There are windows, but it's always been known as the dungeon, because the conference rooms are narrow and the light is poor.

Lucy thinks she can convince her leaders (and me) to redesign this empty space for the three scrum teams. It will solve the noise problem, but more importantly, it will give both of our teams a way to try a new work environment. And of course, give the facilities team a chance to interact with the scrum teams in person, and get quick feedback on how things are working. This work space could be a model for future space in other parts of WL, if we create more scrum teams.

I'm stunned and energized. I ask her if she is sure she won't get fired for this idea. She assures me there are never any guarantees, but she has energy for this change at WL. She wants to be a part of it, and this first move feels like a great way to do it. I tell Lucy it's a great idea, and ask how I can help her or her team.

Lucy laughs, saying we'll have this conversation often in the future, but for now, there are some simple ways I can help.

"Come with me to my meeting with my leaders. Let them pepper you with their questions. Give them the most honest answers you can. Let them discover on their own that someone should be working closely with you and your team. I'll volunteer, and then boom, we all win."

"What about the scrum teams? Have you floated the idea past them yet?"

"No. I better talk with my leaders first. I'm okay with talking with Ben about my idea. I'm sure if we get stuck, we can always pull you in."

"Great. This is a top priority for me, Lucy. Both of our areas need some early progress to get some momentum for the difficult work ahead."

"We can probably crash their meeting tomorrow, if you are game."

"Of course."

After ending the call, I'm back to the C² newsletter. Culture this and culture that. It's up to employees to make a difference. Ugh, I'm full up for now. I close the email and leave it for later. Much later.

Reflections

The way we built our work environments is in direct conflict with collaboration.

Creativity is needed to lead this transformation.

People are hungry to be a part of the change.

I need to find the other Lucys in this company, and work with them. I need to learn their perspective on what needs to be done.

RIE-itis

"Joel, it's Vladimir. Do you have a few minutes today to talk? It's, ah, rather urgent. I wanted to talk with you yesterday, but you were out of the office doing your marathon."

"Ironman." I'm not sure why I bother correcting Vladimir or anyone else. "I'm sorry I missed you. I have time right now. What's going on?"

"I got an invitation to an REI, no wait, an RIE. A Rapid Improvement Event. I have been asked to be a sponsor for the work that's going to happen in the, ah, RIE. I will forward it to you. What's going on?"

Wow, what *is* going on? Sounds like Thad is quickly advancing his work in his quest for world domination. I gingerly lean forward in my chair to stretch my back and see Vladimir's email appear.

"Do you want to talk in person?" I ask. Vlad liked that idea, and so we agreed to meet in his office. This feels like one of those moments to slow down in order to go faster. Vladimir is a VP who normally is very pragmatic; if he is upset about something, he needs my attention. If I don't take the time now, I could jeopardize progress in the future. And, I want Vladimir to trust me. So I stop reading the email, and get out of my office. *Go and see; connect people to vision.* Maybe even some reflection will be needed.

My Ironman legs and exhausted body make the slow walk to Vladimir's office, focusing on my intent. This is the guy who recently attended the first sprint review at WL. This was a chance for the first team at WL working differently to show what they got done in two weeks. He asked a mess of questions, and appeared to be doubtful of the concept of an agile team. But he was only curious. He was one of the very few leaders who answered the last-minute invitation to the event, and showed up.

I struggle to focus on my intent to connect with Vladimir, because Thad is so heavy on my mind. Although I didn't receive the email Vlad received, I glanced at the *TO* address and noticed a ton of people on it. Rick was copied but I wasn't. It wasn't supposed to happen this way. I don't think my team has been slow to start, but this situation makes me feel like we are absolutely behind.

Ugh, I can't freak out about this. Maybe I should talk with Karen and make sure she understood the urgency of getting our team up and running. No, that's not right. Thad pushing ahead of us doesn't mean I should feel behind, and push my team faster than their current pace. That's not a good reason to change our approach.

Eve would ask me, *behind on what?* Because it might feel that way, but it's actually one part of this transformation acting on its own. Going off in its own. I'm getting wound up, and if I'm not careful, I will miss a great opportunity with Vladimir. I push Thad out of my thoughts. *Seek to understand.* This meeting is all about Vladimir, and what has him concerned. I need to seek to understand what's going on. Or, not going on. This helps our transformation begin as healthy as it can. Karen's GEKO model comes to mind. Thad is trying to throw a wrench in *get a healthy start*, although he thinks he is doing great things while the rest of us just sit on our hands.

"Well, there's our athlete, missing a little spring in his step. How was your marathon?" Vladimir smiles and gives me a gentle pat on the shoulder. "Were you successful?"

"Thank you, Vladimir. Ironman Boulder was a success."

"That is right; it wasn't just a marathon," Vlad's nodding. "Impressive. I would love for you to regale your experience to me some day; it sounds like a very difficult day."

"Yes, it was a day full of difficult," I smile, and then we dig into the reason we met.

Vladimir asks me what's going on, again. What I know about this event? He didn't see my name on the email, but he assumed I knew something about it. This is the sticky part for me: how to be truthful without making Thad look like a jerk. How to lead through this? It's also a chance to start forming the transformation I envision, one conversation at a time.

I tell Vlad that his email was the first I've seen of the RIE. And that I haven't read the email that he forwarded.

Naturally, his next question was why not, since Thad and I are supposed to be leading the transformation. I share with Vlad that I'm not sure, but I am aware there are many excited people at WL who are anxious to make changes. They want what's best for WL, but sometimes forget about the others directly and indirectly impacted by the work.

Vladimir rolls his eyes and laughs, a complete break of his professional character that I have never seen. I patiently wait for him to finish reacting. I need to understand more of where he is at before I start talking. I can't lead through what I don't know or understand. I haven't even read what this damn email is asking Vladimir to do.

Vlad raises his eyebrows, "What do you *really* want to tell me, Joel? That you have a rogue leader on your transformation team before it's even off the ground?"

"Vlad, what I really want to do is learn more about what is going on, and what is being asked of you."

Vlad insists I read the email, so I open my small tablet to find and read it:

> You have been invited to attend the first Rapid Improvement Event (RIE) associated with Project Learn. This event will occur Tuesday, August 29th – Friday September 1st 2017, with the final report out on Friday September 1st at 1:00pm.

You have been identified as a key stakeholder of this RIE because you are a leader of an area impacted by the RIE. Other stakeholders include those who have subject matter expertise (SME) on the topic, or because you are a leader of those who are SMEs. ATTENDANCE IS MANDATORY FOR THE DURATION OF THE EVENT. If you cannot attend this event, please contact the CIO, Rick.

What is an RIE?

A rapid improvement event is also known as a Kaizen event. It involves key participants who get together to work on improving a narrowly focused topic...

I look up at Vlad after reading, still at a loss for what to say, how to lead.

"After reading the email, and feeling like I was getting scolded like a six-year old, I researched what an RIE is, and then called you."

"I see."

"The look on your face says it all, Joel."

"I'm sorry, I'm trying to take all this in, and determine what is the most important. It feels like the most important thing is to level with you."

"Thank you. My, ah, position as VP affords me confidentiality."

I do not want to do it this way. But I don't want Vlad to feel like he's dishonest. He's not.

"I appreciate your confidentiality, Vladimir. But this transformation is about all of us at WL making the work and the decisions visible to each other. If we have questions about it, people should be free to answer them and the answers should be available to everyone."

"You and I both know Thad is going off the reservation. Look at that email," Vladimir motions to my tablet. "Telling us we're invited, and then saying it's mandatory that we attend a four-day event. Telling us if we can't attend that we should contact Rick. You know, I don't usually think of my title, Joel, but since I've been VP, I haven't been needed for any work lasting four full days. Spare a conference or summit of some sort."

"Yes, he is. In the last week, I have tried to rekindle our alignment, unsuccessfully. In the meantime, you and others are getting messages from him, or having interactions with him that are not unified with me. Beyond that, the interactions and messages often irritate an already precarious situation, or they create new confusion."

"Which begs the question, what are you going to do about it?"

"I think the question is…what are *we* going to do about it? My part is to lead the transformation of WL. I am going to launch it as planned, and invite him along the way for every step. We're going to make our work known and visible to all. We're going to openly communicate and encourage others to do the same."

"That is fair," Vlad nods. "I own the interactions I have with Thad, not you. And, not Rick."

"And, if I had to guess about that email, Rick probably has no idea that his name is being used as a hammer."

"Of course not," Vladimir contemplates. "Nor will he, until someone tells him…"

"Right."

"Normally, no one would tell him. He might never hear about it… or wait, is he in this RIE?" Vlad turns to his screen to read the email again. "Hard to discern from this message. He's copied on it, and used as a threat, but I don't know."

"Vlad, so what matters to you, here?"

"I'm concerned about how this transformation is beginning. We are not off to a healthy start, and Thad is the number one reason why. You have to find a way to keep him from sabotaging this work before it even has a fighting chance to show its value," Vlad sighs. "But, I also hold Rick accountable for this. His hiring decisions are clearly suspect."

So much for GEKO. Although, a healthy start probably doesn't mean a perfect start. I'm pretty sure that based on my work with Eve, progress will include many bumps, and many steps back.

"What else?"

"Nothing else really matters as much, Joel. I believe we need to change. I don't believe we can force it on people like this message does."

Vladimir goes on, saying he can handle managing his time for this event, but many other people on the list probably won't be able to drop everything for a four-day event that they know little about, and cannot connect to the work they are doing. The frustrating part of this is that the RIE is likely all about their work. This email isn't selling it as anything positive. There is an opportunity to not only share the big vision of why we need to change, but also to invite people to contribute to the answer. To be a part of creating the future, where there is less waste, and more value being delivered to the customer, faster than we used to do it.

Vladimir also points out that the event is the week leading up to Labor Day weekend; hardly anyone is in the office then. For the first event, right out of the gate, Thad appears to be making it as foggy and inconvenient as possible. Vlad is absolutely right.

"This thing is being rammed through, Joel." Vladimir is concerned. "You know why, too. He wants to get the quick win, get momentum on his side. He wants to lead the transformation."

I sigh and look out Vlad's window for an answer. The Rockies are there, unchanged. But they are strong in their foundation. I need to use interactions like these to create foundation for change at WL.

"Joel, I don't know how long Thad is going to last, here, but he can create great chaos before he's gone."

"I need your help, Vlad."

"I know. You have it. In fact, I'm going to take this particular issue and own it, if that is acceptable to you."

"Yes, that is great," I smile. I am getting help! And, from Vladimir, who looks to be a dark horse transformation champion in the making.

"You inspired me today, Joel. When you first arrived, my intent was to persuade you to take action on this problem. You've made me reconsider my approach."

"That's good to hear, Vlad. I do care deeply about this work, and about how we are asking people to change."

"Understood," Vlad nods. "Besides, Thad is becoming a force by the day. I can't expect you to wrangle him all on your own."

"Thanks."

"It might not be all bad; it's still possible for Thad to fold into our tribe, and still be a lean expert and leader. But from what I've seen, we have a negative force on our hands. It's going to take many of us united to keep Thad from destroying all that we've worked so hard to build at WL."

Reflections

Rapid Improvement Events at WL are not off to a good start.

Change doesn't work well when people are forced into it.

Leading through problems requires ridiculous amounts of listening.

Vladimir is a dark horse transformation champion in the making.

Executive Calendar

"Everyone, please meet Joel. He's going to be teaching us about how to get the waste out of our calendars." Rick introduces me to a select, group of senior executives. We've all seen each other before in different work environments, but I'm fairly certain I know their names and they did not know mine.

"You were all so impressed with the changes I've made to my calendar, and subsequently, my life, that you requested to learn how to apply it to your lives as well. I don't have all the answers, and Joel probably has a few more than me, but let's have the conversation. Let's explore how we might work differently."

As a result of his work on his calendar with me, and a focus on value, his blood pressure readings have dropped enough that he may soon no longer need medication for it, and he's able to spend more time with his ailing, 85 year-old father. He also feels he has far better focus at work. Mostly, because there is much less noise rattling around in his head.

"Joel, can you tell the group more about this phenomenon?"

I share the concept of the Zeigarnik effect, right from the Personal Kanban book. This soviet psychologist studied the human brain's need for closure. In fact, our brains are very preoccupied with loose ends, because they seek out process and meaning. This is why Rick became so overwhelmed with life and work. He has many things happening at once, that a loop of worry began to form. This is the Zeigarnik effect. It's not important to remember the name or the study, just know that when our brains seek meaning and closure to loose ends, and we have no means to pull those loose ends together, we ruminate over them. Over and over again. The room groans in appreciation. They all have felt this, and now I feel connection with the group. I was hoping this would happen. Now, to keep them engaged.

I encourage Rick to continue with his story. Rick learned to see where in his interactions and meetings he gets value, and where he gives value. He admits looking for value is not magic; his job isn't easier as a result. In fact, leading WL through massive change for the better is more difficult than ever. Focusing on value offers clarity. If he could distill it down to one convincing thought it would be that he has confidence he is working on the right thing, at the right time. This is because he is no longer making unconscious decisions, which he believes led to having an unconscious day. Maybe an unconscious week, month, year, or several years. This comment caused some leaders to shift uncomfortably in their seats. Rick acknowledged their discomfort and invited them to just let his experience sink in, and not judge it in this moment.

Wisely, he mentioned how this change impacted his interactions with his assistant. He had to train his assistant on his new approach; of course she accepted the challenge. The result is that she loves it. With Rick focused on value and intent, she is doing less speculating about what matters to him, and their conversations are more efficient. In fact, Rick looks forward to his time with his assistant, because they are both energized by this new way of working.

Someone asked him how long it took to train his assistant, to explain everything to her. Rick assured the leader it was a 30-minute conversation. Since they talk daily anyway, there is opportunity for his assistant to validate some of her decisions.

Overall, this change has been one of the best he's made in a very long time. He feels happier and has more energy. He reiterates that he doesn't have all the answers, and still has more changes to make in his life. Next up is to get back in the gym. Some of his peers nod their heads, as if they know exactly what he is talking about.

"So, as we embrace the vision to *grow happy, grow strong,* we could live out this mantra ourselves, through our calendars. This might be a way for some of us, or maybe all of us, to show how we are working differently. We may not be on an agile team, but we can live out agile principles."

There is some short discussion about agile leader training, maybe adding in an agile leader coach.

"I have you to thank for that, Joel. Thanks for being the first one. We need people to go first." Rick tells the group the purpose of the meeting is to learn at a high level, the concepts of value, waste and visual management.

"Joel, here is practicing coaching and mentoring, part of lean leader standard work, and being an agile leader. I am going to learn from Joel. This means Joel will get us up and running with teaching us the foundation, and I will be able to do the follow up with this group. I don't feel entirely prepared for that, but Joel, here, believes I am ready."

I thank Rick and smile. This topic with this group makes me nervous. I'm a mid-level tech director telling senior executives they have waste in their day. This could end badly for me, or even for Rick. I sense Rick feels the same way, and is very engaged in my every word. It's neither comfortable nor uncomfortable.

"I can't recall the words you used to describe the awareness I had of the noise in my head, Joel."

"Without getting too deep, it's metacognition. Knowing about knowing. Rick, you became self-aware that you were swimming in information, responsibilities and tasks. And, you made a choice to do something about it."

We only have an hour together, so I planned to give them about 30 minutes of context, and 30 minutes of looking at their own calendars. Eve called it *hands on* work.

I begin teaching the group that the goal of the lean methodology is to provide the best value to the customer with as little waste as possible. Waste is a huge topic that I recommended the group research, just like I had asked Rick to do a while back.

"The tricky part for a leader's calendar is figuring out what is a value-add for the customer. Our current culture at WL has a different idea of what this is than what the lean and agile methodologies offer. For most leaders, there are two kinds of waste: the things we *must do* that do not add customer value, and things that *we think we have to do* that add no customer value."

Rick asks, "Can you please share the examples with this group that you shared with me?"

"Yes," I pause to make sure I have their attention, just like Eve would do. "It's important to remember that lean and agile methodologies uphold that people have the best intentions. People are never waste. So when I talk about waste, it's something that is broken in the process or in the environment, not the person."

The group nods and waits for the examples, appearing to grasp what I am saying.

"An example of waste that we must do, but does not add customer value is preparing and submitting your budget numbers for the next year. A leader has to do that to keep the machine running, but the customer really doesn't care if this happens."

The executives nod and make a few small comments and jokes about budget submissions being some of their favorite tasks.

"So, that is a non-value-add, or waste. But it's something we have to do. There may be a better way to do it, to improve it, but it still needs to happen."

I'm writing on the whiteboard in the front of the room; the leaders are all taking notes either on their tablet or in a notebook. No one has their nose in their laptop, pretending that something else is more important.

"The second example is of a leader who has 1x1 status meetings with each member of the team. It might seem like these meetings are necessary, but they are not. And, the customer could care less if these meetings happen. This is an example of non-value-add that the leader doesn't have to do."

As expected, the same question arises that did when I shared this topic with Rick: how does a leader know what's going on? How will the leader keep in contact with the team?

I tell the group that an agile leader doesn't stay in the office and make people come to them. They practice *go & see*, a lean leader principle. The agile leader serves teams by getting impediments out of the way, and making space for teams to find opportunities and innovate. In order to do this, leaders go to where the work is, which is where value is added. This is called the *gemba*.

"Yes," Rick adds, "the gemba is the actual place where value is added. Don't worry about using the Japanese word for it. If you do, great. If not, that's okay too."

I love that he's at least trying to learn these concepts. I need all of the champions I can get, even if they are only grabbing onto part of the change.

The group is stunned. They start talking among themselves. I wait for them.

Rick wrangles the group back to the topic, "We realize that customer value and gemba are huge topics, and today we are just scratching the surface. Joel has wisely coached me that it's better to begin practicing, than to talk about it and never go forward. Just like we heard at the leader summit earlier this year, this is a different way of working. We have been asked to be bold, and be willing to make mistakes, because this is not easy."

"So let's practice," I invite the group to open their laptops. We begin the exercise of looking for value-add, and non-value-add meetings, and for waste. I remind them one more time that people are not waste, but processes and events can create it, even those with the best intentions.

There is a lively discussion as the group explores their day with a different perspective. This is the moment I intended to happen in this meeting, and so it is. What they do with it after today is up to them.

They find one meeting that doesn't have value to any of them. It's with a metrics group. Apparently, this group gets the same numbers from two other reports. No one understood why they still had to attend this meeting. Another discussion ensues.

"The first time Joel had me do this, it was very difficult. And very exciting," Rick says.

The group was chatting and making comments about how hard it was to look at even one day of their calendar. One leader said he wanted to spend more time working on his calendar, *but he had no time*. The group laughed, all feeling the same sentiment.

I acknowledge the comment, and share that our corporate environment has many facets that can change, and this work could be one of them. I believe changing how we use our calendars would be a huge step toward changing the environment at WL, but I can't say that directly to this group. I wasn't brought in to persuade anyone. Only to teach them. Even that is a stretch; I'm a director in tech, not a gemba coach.

"We've been programmed to have meetings for everything, and most of the time, teams have to leave their work to come and talk with us," Rick says.

"Yes," I say. "Our current environment makes it look like customer value is added in the leader's office, but usually, only teams actually add customer value. Leaders are an important part of the process, helping remove barriers for teams, and setting the direction they are working."

I can't believe I just said that to a group of senior leaders. They look very deep in thought. Or, they are thinking of the different ways to fire me in a slow and painful way. I might as well push a little further. We have time and this may be my one shot with them as an audience.

"An agile leader works relentlessly to connect people to vision. It's one of the most important things a leader can do, because when people hear the vision over and over, two things happen. First, they trust the leader more, because each time they interact with this person, they get the same message and intent. Second, they have a strong grasp of the vision and can use it with their interactions in the enterprise."

Something in Rick's body language changed. I think that last concept made Rick nervous, or he's worried about our time constraint. Maybe he's just not ready to go there with this group.

"Back to the calendars. Joel has stopped having his 1x1 meetings, except for career development purposes. Since he spends enough time with his teams every day, he knows what is going on." Rick looks proud of me, more than I have ever noticed.

"Rick makes this change sound easy, but the transition to it wasn't. In fact, I know more now about what's happening than I ever did in the parade of 1x1 meetings I had in my office. First, I was trained on how to go to my teams and see their work. Then, I prepared my teams by telling them what I was going to begin practicing, and why. Only then did I begin practicing walking the gemba."

I tell the group that it's not about checking on employees, but engaging them at the place where the value is added. The concept originated on the manufacturing floor, but can be adjusted for knowledge workers. Don't plan to take a lot of notes or input things in a chart. This is not a chance for judgement, but curiosity. It's very hard to change to this mode, because we've been conditioned to drive for results, not to be curious. An easy way to stay focused on curiosity is by using three simple but powerful questions. A leader, or anyone for that matter, can easily learn about someone's work using them:

1. What are you trying to do?
2. What have you tried so far?

3.   How will you know it's working?

"And, there is more than one way to ask the questions," Rick adds. "But we are just about out of time for today."

Everyone looks at their phones or their watch, and then begins to pack up. They express to Rick the wish for more time together on this topic. That's a win.

"Joel, thank you for making time to be with us and for all of your good work." Rick puts his hand on my shoulder. "This was great. We have a lot to think about, now. We'll be sure and call you back if we have any questions, or want to explore the topic any further."

As I leave the room, I'm sweaty and slightly exhausted, but overall very pleased with how I delivered this concept. I had just one hour to get their interest and instill a hunger for more. I think I succeeded. And, I think Rick has confidence to do further work with this group. Okay, maybe not quite that confident…

"Joel, do you have anything right now?" Rick stops me in the hallway.

"Uh, I was going to visit Vijay's team, but I can hold off for a few minutes. I will just text him that I'm going to be a little late." Actually, I text Vijay that I can't make it at all and why. Vijay knows exactly how these conversations play out; 10 minutes easily becomes 45 minutes, or more.

Rick nods, "Let's duck into this conference room."

I can only imagine what this conversation is going to entail. I just delivered ground-breaking leadership messages to a group of Rick's peers. I'm not a coach or a trainer, so I'm positive I messed up something.

Maybe it's not even about that. Maybe he's going to thank me for pushing back on him to involve Vijay's teams with the big data cyber security teams. We just dove into that work, and never really debriefed that original meeting. Who am I kidding? Rick has moved on from that. Meanwhile, the team is knocking it out of the park, working so well on the issues, and making huge progress. The small director team is even impressed with their progress. So much that they are starting to not attend the daily meetings. I'd love to talk with Rick about that, because I'm sure he doesn't know about it, but now is not the time. Besides, that's something for Vijay and I to attend to. Soon.

"You did a great job in there, Joel. I'm very pleased with what we accomplished in just one hour. Thank you again."

I wait for the shoe to drop.

"What I want to discuss is unrelated to today's event. I have some feedback for you as it relates to how you're working with Thad. Well, it's Thad's feedback for you."

"What's on your mind?" I can't imagine...

Can't imagine what Thad would have come to Rick about. Can't imagine how he packaged it, and twisted the truth. Can't imagine why Rick is delivering Thad's feedback for me. Since when is the CIO the messenger for one of his direct reports' messages to deliver to one of *his* direct reports? No, I can't imagine how any of this is going to be progress forward in my agile leader journey.

"Thad has brought to my attention that you're not aligning very well with his work. He's also very concerned at what he described as you being...on edge."

I wait, knowing better than to say anything at this moment. There is more coming...

"Thad said that you *came at him* and *attacked him personally* for not communicating with you about his work," Rick sighs. "He also said that you are acting sneaky, and keeping secrets from him."

Wow. Basically, Thad is accusing me of everything he has done. Except the personal attack garbage. That's pure narcissistic personality disorder speak for *I don't like how you called me on the carpet.*

"There's more, Joel. Thad saw you as resistant to lean and lean thinking, and that this is creating a bottleneck for him. I might not get this straight, but I think the words he used were that you have a *myopic agile view of the world.* He believes you will sabotage the gains he is beginning to make in our operations space." Rick sighs once again. "Over the course of several interactions, Thad has come to believe that you mean well, but you are not ready for the challenges of the transformation work. He is concerned that you're overwhelmed with your responsibilities, Joel. He believes this is the reason you are acting this way. He even said he felt sorry for you when you lashed out at him. And because he's a man of empathy, he offered to take on some of your work, temporarily, to give you a chance to get your bearings."

I nod my head slowly, then regret that I'm nodding my head. Rick could take this as agreement. Ugh. There is no good way to stand here and listen to this BS. I am the bottleneck?!

"Joel, I know you pretty well. These things Thad is saying about you, they don't sound like you at all," Rick offers up the classic feedback language. Everyone at WL has heard it before. No doubt about it, I'm getting nailed for this. So much for being a few moments late to Vijay's team area.

"The worst part is that some senior leaders caught wind of your misalignment with Thad, and now *they* are concerned. They have voiced these concerns to me. Some of them were just in our session. I can't tell you who, of course. Just know that they know this is a problem. They don't know Thad or you, but what they see is two WL leaders detracting from the *grow happy, grow strong* vision," Rick sighs again. "Joel, I can't have situations like this at the start of a transformation. It's only going to hold us back from our full potential."

Rick stops. It's time for me to say something. Eve has taught me that an agile leader should always be curious. Whether the situation is an opportunity or a challenge, I should be asking a question. So, given that Thad managed to shred me without even being present, what question should I be asking right now? What do I want to learn?

Other than the obvious, *are you going to fire me?* I want to know if Rick is going to let me handle this on my own, or is he going to get involved. Actually, I want to know if Rick wants to hear my side of the story. I want to know what Rick actually thinks of this story. The way he delivered his feedback wasn't very neutral. It seems like he's bought in to all that Thad said. Hell, the fact that Rick is giving me Thad's feedback without him here tells me something about this whole equation.

A CIO doing someone else's leader job. Why is that? Because he believes all Thad said and wants to personally handle the mythical crap storm I've created? Or, because he suspects there is another side of the story and he wants to hear it from me. If that's the case, he sure hasn't let onto it. This scene is a big, fat, mysterious mess.

"The easy thing for me to do would be to remove you from this work, and give it to Thad to lead."

So much for wanting to hear my side of the story.

"But I'm not sure that is the right move. You had some very good ideas, Joel," Rick smiles. "We have talked in the past about Thad, and that he has a lot of great things to offer our organization. I shared my fear that many leaders here won't be able to gel with him. I put my trust in you because you are one of the few leaders I have who is approachable, and seems to be able to cut through the stiffness and dogma. Now it seems that you're not able to work with him, either."

"O. M. G.," Meryll is pacing in my office. Jack and I are seated. Our usual coping postures.

"Then what?" Jack asks.

"He said he could ask me for my side of the story, and that he knew my side of the story would be dramatically different than Thad's story. But he didn't want to spend the time doing that. He wanted to move on, and look forward."

Meryll's mouth is agape.

"Since he said he put his trust in me when Thad moved to operations, I asked him if he still trusted me. He said yes, but he said Thad's concerns have put him in the *hot seat*, and that I need to quickly turn things around with Thad."

Jack tries to be positive, "At least he still trusts you. After all, you are the Agile Transformation Sultan."

"I've morphed to just TS, Jack. Transformation Sultan. It's so much bigger than agile."

"Roger that."

"Jack, you don't get it," Meryll stops pacing. "Thad is controlling this whole thing! It has nothing to do with Rick trusting Joel. That was a very easy thing for Rick to say, because it has nothing to do with the problem, which is Thad."

"Ugh," Jack puts his head in his hands.

I just smile.

"Thad is becoming this separate entity within WL. Somehow, he's managed to manipulate the guy who hired him, and apparently, other senior leaders. He's doing whatever he wants, saying whatever he wants."

"Exactly right, Meryll," I say.

"The day a VP can move our CIO up and down like a lever…I never imagined it," Jack sighs.

"It's here," Meryll paces again. "What about your coach? It sounded like Rick was not happy with your coach."

"No," Jack playfully corrects her, "Rick said that *Thad said* Joel was misguided."

Meryll rolls her eyes, "Right. Thad could coach all of us on how to be successful, couldn't he?"

"Making our way back to the big problem," Jack says, "our CIO is being manipulated by a VP."

I tell them I agree, but I don't think Rick is oblivious to Thad's strategy. I don't think he likes it. I think he is putting up with it, to use Thad for his expertise.

Jack is thoughtful, "So, he's looking to you to *gel with the leaders* and get aligned. Sort of, *please put up with this nutjob so we can get the benefits of his knowledge.* Very flawed, but I see it. That has to be it."

I laugh, "Even better, how about: *I know I hired an ass who is smashing into everything he touches, but I can't fire him yet because that would make me look bad. You're one of the few people who can work with this guy, so won't you please figure out how to do it better?*"

"Right on!" Jack high fives me. "Well, either that, or it's all a front for the zombie apocalypse."

We reminisce about the leader summit earlier in the year, when we were reminded of a mistake Rick made in 1993. That event, and all its grandiose encouragement to *be bold* and *be brave enough to make mistakes*, proved how difficult it is for WL leaders to admit a poor decision. Here we are, with a CIO who would rather look the other way than admit a mistake. We all wonder how big the path of destruction would have to be for him to admit the mistake. We may never know. That would be the simple way out.

"What a joke," Meryll huffs. "So what did you say to him after all that? Did you say anything? I don't know how you did it, Joel. I would have been too stressed out to say anything positive or helpful."

"Yeah Joel, how did you channel your inner Eve?" Jack holds his palms open in front of his chest.

"Well, I was right there with you, Meryll. It took all I had to listen to Rick spew this BS about me. I'm sure my face showed some sign of stress. I'm human and a professional," I sigh. "Thad used the behaviors I called him on a while ago, and turned them around on me. It was a moment of insanity created by a narcissist."

"Well, at least you didn't blow your stack," Jack offers. "You would have got fired for that. Maybe Thad was banking on it."

"Or," Meryll looks out the window, "he's hoping you won't be able to stand the heat, and you'll quit. He knows Rick holds you in high regard, so he had to spread bad things about you to other leaders. He's building a case where he can."

"Sickening, but you're right, Meryll," Jack slouches in his chair. Even the optimist has his limits. "Honestly, Joel, Thad's got it in for you. And, this is a big message to anyone else who wants to *align* with him."

"It would have been easy to tell Rick that I would talk to Thad and straighten it out, but I didn't. At this point, I am aware that we both know there is no *straightening things out* with Thad. This is when I channel my inner Eve. Man, if there was one time I wished she could be with me at work, this would have been it. Not to tell Rick he's messed up, but to give me the right thoughts for the moment. I was really stuck on what to say or do."

Meryll sits down in my other guest chair, "I get it, Joel. And, I know exactly what you did."

"You do?" Jack and I ask at the same time.

She grins, "You thanked him for the feedback."

Jack shoots a glance at me to validate.

My raised eyebrows are all the signal he needs. He pumps his fists in the air, "Yes!"

"Brilliant, Joel," Meryll is still smiling.

"After all, it was just feedback," I shrug.

"You can't just leave us with that, Joel. Then what did Rick say to you?"

"He said you're welcome, and he had a faint smile."

I wait for my friends. One of them is going to figure this out.

It only takes about twenty seconds before Meryll throws her head back and says, "God, Joel. He knows he is screwed! Ugh, why can't he just *own it* already?"

"We all know why," Jack is caught up.

And we do. We may have heard about one senior leader's mistake from years ago. We may have heard the message to be brave, and be willing to make mistakes. Now WL has a manipulative control freak to replace the one that was recently fired. Only this one is disguised as a lean expert and a lean leader. No WL senior leader is going to admit to making a mistake of hiring the wrong VP. Certainly not the VP who was specifically brought in to help lead a massive transformation. The only thing people will learn from this situation is how to manipulate it further, or how to avoid it all together.

So the hiring leader will buy time by dancing around him, pressuring others to adjust to his style, and pressuring others to work with him, all in the name of alignment. Then others are made to be the problem not him. Meanwhile, the rogue leader continues his path of destruction.

"I hate sounding like a negative person, but there's gonna be a trail of bodies behind this," Meryll warns. "It seems impossible for you to not be one of them. I will do whatever I can to not have you be one of them, although I don't know how to help you. But I'm hopeful that Rick can help you."

I shake my head no, "Rick will only be able to do so much. As long as he's going to hold onto his hiring decision being the right thing for WL, I have limited air cover. In fact, that conversation we had might be the only thing he will do. I'm exposed right now, and will continue to be indefinitely. But thanks for having my back, Meryll. I will ask you and Jack, if there is some way for you to help. Even if it's to dig my grave."

Meryll rolls her eyes, "Ugh. Stop it, Joel. I just hate that you're in this position."

"Yep," Jack sighs. "Let's say Rick actually did admit a mistake, and said he was walking the talk about leading differently and continuous learning, his peers and senior leaders are not ready for that. They would squash him. And he knows it. So this is all going under the rug, and you with it. No mistake made here. And the only thing anyone is going to learn from hiring Thad is how to spin, manipulate, and avoid."

"We all know how to play that game," Meryll mumbles. "All too well."

"Okay," I try to perk up. "I'm going to go forth with a strong, customer-focused transformation offense. The team's work will speak for itself. I will give them my best sponsorship and change leadership, otherwise known as *stakeholdering*. And, I've got my resume in decent shape."

Jack high fives me again, "You're gonna own it like a boss."

Reflections

Senior leaders hold promise for me that WL might be able to change.

It's uncomfortable to teach others that customer value isn't delivered in a leader's office.

I am put in a place of friction because a leader can't admit he made a mistake. Is there hope to still learn from this one?

WL pressures leaders to have all-or-nothing thinking with who we hire.

My interaction with Rick went better than planned because I listened to understand.

Thad is not a bottleneck to the transformation; he is a threat to it.

The Unlikely Coach

My walk to J&L's Café for my next meeting with Eve is noisy. Plus, we haven't talked since Ironman Boulder. After all, I'm wearing my new Ironman Boulder finisher jacket. I'm sure she's going to want my reflection on that as well. I try to think of my intent for our time today: what do I really want from our meeting? There is so much dysfunction right now, I'm not sure if my next move is the right one. Yet asking Eve for validation of my next right move is too shallow of a purpose. She won't accept that as a coaching need.

Eve has trained me to approach situations with curiosity and an intent to understand. I need to be thinking *What question should I be asking right now?* She wants me to coach and mentor others to do the same. I used to think I wasn't ready to do that, but after practicing a few times, I did well. In fact, it's a little easier to help someone else be curious and seek to understand, than it is regarding my own work and leadership.

So if I could ask just one question of Eve regarding this swirl of chaos, what would it be? I guess it comes down to... *What should I be doing right now?* No, that doesn't seem right. Too general. Maybe, *How do I lead during chaos?* I like that the question includes something about leadership but it's still too basic. I want to show Eve I'm not just complaining about WL. How about, *How do I lead with focus when no one else has it?* I'm at the entrance to the café, so I guess that's what I will go with.

I'm waiting in line at the café, when Eve comes up behind me. She greets me, and tells me we are going to do something different today.

I smile at her, "Of course we are. Why would I expect our time together to be a garden variety coaching session?"

Eve ignores my comment, "It's time you practice again, Joel."

After we get our drinks, Eve leads us to the back office of the café. This is where I last helped the owners, Lilly and Jessica, build their personal Kanban boards. I wasn't at all ready to coach them, but Eve thought I was, so she pushed me into helping them. They loved it. At least, at the time they did. Their next assignment was to build a visual management board for their new restaurant idea. Eve probably wants me to do some follow up work with them on the board.

Eve and I find the owners talking in front of a new board. It must be their restaurant board. We greet each other, and then Eve gets down to business.

"The purpose of our meeting with you has two parts: one for Joel to follow up with you on your personal Kanban boards, and your new restaurant board. We will revisit limiting work in process, value, all of the great work you've been doing on your own. The second part of our time is to help build your entrepreneurial spirit. This will be great for Joel to practice a new concept he learned."

Entrepreneurial spirit? Practice a new concept? My heart rate is up and I'm already feeling warm. A little advance notice for this stuff would be nice. At this very moment, I'd say agile leadership is very overrated.

Eve asks me to revisit the personal Kanban boards each owner created, so I'm on. Man, now what? Intent. I should begin with intent... We should review the reason they have a personal Kanban board. Seems like a good place to begin. We walk over to their personal Kanban boards. I'm sure I'm going to mess this up.

"Uh, does anyone remember what the purpose of your personal Kanban is?"

"It's a way for each of us to visualize, understand, and improve our work," Jessica says. "It makes it easier to see where value is."

"Great." I build on Jessica's answer, stating that the only two rules for the board are to make all the work visible, and limit how much you have in progress. They tell me they still love the simplicity of the board, and that it's helped them both be far more productive, and make connections to their work.

We discuss value, and how the personal Kanban shows the flow of value for an individual. Flow, in this case, is the flow of value: from an idea to the time that idea becomes something to help the end customer.

"What have you learned from working this way?" I ask. "I want to hear what you've learned. The good and the bad. Well, maybe not bad, but…uncomfortable."

Lilly and Jessica smile at each other, and then Lilly says, "We both really struggled with limiting how much we can do in a day. We always knew we planned to get more done than we actually could, and personal Kanban proved it."

"The busier we get with our restaurant planning, the more I have used this board. I used to have so many things flying around in my head, that it was difficult to fall asleep. I was either afraid I would forget something, or I wasn't sure what I was going to do first in the morning."

Jessica and Lilly share that they each review their personal Kanban boards before leaving for the day. And, it's the first place they go when they return the next morning. They learned that sometimes a good night's sleep changes their perspective on what is the most important work for the day. Since the work to be done is set up as *options* rather than a list, it's very easy to change priorities.

I explain to the owners that what they are experiencing is the sweet spot of personal Kanban. Big projects, life changes, even parenting responsibilities create many bits of unrelated information and tasks in our heads. They swirl around during the day, clouding our focus and our decision making. They continue to swirl around later, holding us back from relaxing, even sleeping.

On top of that, we have constant interruptions all day long. Many of them are digital, but for café owners, the stops and starts of concentration come from all angles. Our focus is impaired by all of the interruptions, costing us time and ultimately, money. Personal Kanban gives us that focus. With incredible focus, we can keep learning and adapting on the best way to tackle the work in front of us.

I can't believe those words are coming out of my mouth. I didn't even know what Kanban or personal Kanban was a few months ago. Now, I'm coaching two business owners on how to use them, so they can launch a new restaurant. I know Eve would stop me if I was saying something that was really wrong.

"The fancy name for what you're feeling is existential overhead," I continue. "But most people call it clutter or even chaos."

"You got that right," Lilly agrees. "It's like a ten-pound weight on top of your brain."

"Personal Kanban takes these pieces of uncoordinated information and tasks, and puts them in a framework for systemic understanding. This is your value stream, with your options clearly in front of you." I still can't believe those words came from me.

"Great work, Joel," Eve says.

"Can we tell you something else?" Jessica asks.

"Of course," I say.

"One of our favorite things about using personal Kanban, and our restaurant Kanban board is when we get to move something to done," Lilly grins. "I never imagined such a simple action could feel so good."

"Total brain candy," Jessica gushes. "Almost as good as our medium roast."

"There's a dopamine rush," I say.

"Huh?" Lilly and Jessica say at the same time. Eve is just smiling, as if she believes I know what I'm doing. I sure don't

"Uh, neuroscience research found that when we humans finish tasks, we get a dopamine rush. We feel really good. I agree with that," I say. "I love seeing the lawn after I just cut it, or a pile of strawberries that I just cleaned. And of course, when I move something to *Done* on my own personal Kanban board, I feel great."

"I totally get that," Lilly smiles. "It's so fun to move stuff to done. Sometimes we fight over it!"

Then I ask the owners how they are progressing with their visual management board for the restaurant. They walk us over to the board, and then give me a chance to read it. That's when I realize I'm not supposed to just read it; the board is supposed to encourage interaction. I ask them to walk the board with us.

"Please show us how you use your board."

The owners begin with what is in *Doing*, and finishing with the items in their backlog. It seems to me that they have a good understanding of their board and their work. I look at Eve, hoping to see some sort of direction on what I should do next. She is of no help. Of course not. This is mine to figure out.

I should be asking a question right now, I know that. Since I'm at their gemba, maybe I should practice my questions. I look at one of the tasks in *Doing: write job descriptions*. I ask about this one, "Is there anything getting in your way?"

It turns out the job descriptions task is a hot button for the owners. They want to write descriptions for themselves, so they can at least in theory, divide the work in a formal way. Also, they want to write job descriptions for the managers and the employees. It sounds like they have a lot going on for just this one task.

"Yeah, that one should be made smaller," Lilly says, "because there is no way we will write all of the job descriptions today."

"So, we should make a sticky note for each of these tasks?" Jessica asks.

I nod, "You need to make your work small enough to make sense. Maybe that's one for each job, or maybe it's broken up some other way."

The owners talk among themselves, remove the sticky from *Doing*, and then quickly write more tasks on sticky notes, and put them in *To Do*. They move two tasks back into Doing. One task is to write a job description for Lilly, and one is to write a job description for Jessica. I ask them to take another look at the board, and tell me if it feels better. They tell me it does. Then, they begin looking at the other tasks they have in To Do, and more discussion ensues. They found other work on their board that could be broken down further, to make it easier to see and understand. Eve motions for me to step back from them, and allow them to work.

"Well done, Joel. You asked some good questions. An improvement would be to ask what's getting in the way. Don't pick a random task and ask about it to look interested. That is whimsy, and at best, guessing."

"Ugh."

"It's okay, Joel. It's why we practice. People tend to hang on every word that leaders and coaches say. Even things just in passing suddenly become a high priority or hot topic."

"Yes, I know that well, Eve. Then later, the leader wonders why the focus on that item."

"Right. You get this. Be curious with intent, and stick with your three powerful questions. No executive whimsy needed."

We rest for a moment.

"They are fast learners. I'm so impressed with how they just…get it."

"Like all of us, they are hungry for clarity and focus."

Lilly and Jessica stop, and then apologize for getting so involved in their work. I tell them not to worry, that interacting with the board should always spur lots of conversation. We finish by me asking them if they have any impediments to the work we just discussed.

Lilly looks at Eve and asks, "Can we move to the second topic now? We think it's an impediment."

Eve shrugs, "Up to you."

Lilly and Jessica say they are ready, so Eve encourages them to tell me about their impediment.

Lilly begins, "It might seem strange, but we are moving way faster than we thought we would. It's a good thing, but it's also a bad thing. By organizing ourselves with our personal Kanbans, and our Kanban for our restaurant, we are so much more efficient. We never imagined in our wildest dreams that we'd be this far along by now. It's incredible, and it's scary."

I remain silent, and wait for Lilly to share the impediment.

"We find ourselves completely freaked out about being so close to launching a new business," Lilly says. "It still seems impossible."

Bingo.

Opening their restaurant is like my Ironman Boulder hill. The big one, at miles 13.1 and 26 of the marathon. The one I couldn't imagine myself doing, let alone run up it both times, the last one while holding a bag of cheese curls.

"So, let me get this straight," Eve speaks slowly. "Using Kanban and personal Kanban has accelerated your work on your new restaurant beyond what you ever expected."

Jessica agrees, "We know it sounds strange. We've been talking about this for so very long, and now it's real. Very real."

"And with it being right in front of us, we got freaked out. Well, we *are* freaked out," Lilly says.

Eve asks them what they are worried about. What is causing the stress?

"Mostly, that we've been talking about it for so long, but it was far off in the distance. Now, we've covered so much ground so quickly, it's no longer far away. It's right here in front of us. And now that it's right here in front of us, it seems impossible to own both a café and a restaurant."

"Impossible, unlikely, or just…really hard?" I ask. So simple and impactful.

"That's a great question, Joel," Jessica says. Eve nods her head in agreement.

"You want us to pick one, but it feels like all three of them," Lilly sighs.

"Uh, well, there is a difference between the three. You're thinking it's impossible to open a restaurant, but it's probably not. You've got the board to prove it. You've done the work; you have mapped it out and you can see the flow of value. You work together as a team very well. The focus of the board, the flow of value, the communication around it…this is great stuff."

"So, it's unlikely," Lilly's wheels are turning. "It's risky. So many new restaurants fail. So many small businesses grow too fast, and then have to pull back, or go out of business altogether. We see it all the time in this market. That freaks us out."

Jessica perks up, "Now we are getting somewhere. I think."

I tell the owners that what they are doing is very hard work. And, there are no guarantees that they will be successful. This is what freaks us out. We know success is unlikely. This is why most people back away from this kind of goal. They know the risks. They all know someone who tried and failed. Worse yet, it might have been them who failed. They know success is unlikely.

"So, you're giving us motivation instead of another tool," Lilly announces.

"Correct," I make a side glance to Eve, who is acting as if I'm doing a great job. I don't think I am. I need to step it up. I hope I can step it up. "But I'm not going to stand here and say with false encouragement, *you got this!* This isn't about me liking you and your café, so therefore I can motivate you to step past your fears. This is motivation because you are ready," I pause. "You two had a dream or an idea a while ago to open a restaurant, right?"

They agree.

"You believed in each other enough to dream together. Plus, your current success gives you confidence that you know how to run a thriving business. I can attest to that, as I'm usually in some sort of line for my medium roast. You've already accomplished the unlikely. That is more than many business owners can say."

It's also more than most transformation leaders or sponsors can say…

"So, people who can do something difficult might not want to try something that's unlikely," Jessica says.

"Right," Lilly nods, "because they're not willing, not knowledgeable, or not interested in how uncomfortable it can be to go after the unlikely. We have to be willing to take stock in our board, all of the mapping out of our launch, and our financials. Those are the knowns; we have to forge into the unknown."

I hold up my hands, "You nailed it, Lilly. What you are doing is not impossible, but it is unlikely. So the question to ask yourselves is, *Are we confident enough in what we know and have, to take that step into the unknown?*"

"And, there is a ton of really hard work that leads up to this point. We were able to buzz through that quickly using Kanban. Even if it wasn't easy, it felt pretty safe. Writing a job description doesn't freak us out. So, with most of the hard work out of the way, all that's left is the stuff that pushes us into…unlikely."

Huh. Their work is so much like mine in this respect. The entrepreneurial part of it and the unknown part. Most of us at WL might be willing to do the difficult work, but few of us are willing to do work that has no guarantee of success. No wonder I feel so uneasy every day. I'm leading a team and myself to do the unlikely. There's a lot of risk with it, and the environment is nearly hostile for this type of work.

Lilly and Jessica are now facing each other, grinning. I think this means our conversation helped.

"Jessica, we got this," Lilly gently grabs Jessica's shoulders. "We are freaking sharks. We don't care what day of the week it is. We swim around, look for opportunity, and then BITE."

They high five each other. I'm sweating, but I'm feeling good.

"We totally got this!" Jessica proclaims.

"We are in this business because we want to be. We love the rush, we love the connections. We aren't really afraid of this!" Lilly is practically jumping up and down.

"Stay focused, keep our work in front of us, limit how much we do, and we will be all right." Jessica sighs. "I feel so much better. I'm still freaked out, but I get what it is now."

They both turn to me, and overflow with gratitude for my help. They ask me how I know so much about doing the unlikely. I look to Eve, and she encourages me to share with them. I know if I don't tell them, she will. So I should own this like the change leader I am.

I tell them I am no expert, but I have two new experiences in doing something very unlikely. One of them is leading the transformation of WL's technology areas, so that the company can work differently. We want people to change how they think about their work, and how they do it, similar to what you two have done with your personal Kanban and restaurant Kanban boards.

Leading this change has been an overwhelming experience, and I'm just beginning. I know the work is not impossible, because so many companies have done it successfully before WL. There are coaches, conferences, books, and all kinds of training for leaders, teams, and individuals to access so they can change.

I believe it's unlikely to work because many people must do it for it to be successful. And, it's a mindset shift for people in how they do their work. These are two huge things that make it unlikely that we will succeed. There are other reasons, but these two send the change over the top. It feels impossible. I don't know if we will make it, Eve doesn't know if we will make it, but I am going to give it my best. I'm also going to ask those who work with me to do the same.

"So, you are leading a change that seems unlikely to you, and to a bunch of people," Lilly says.

"Yes. My team and I will have to remind them over and over again of our vision and support. That's the thing; most people learning of how we want to change WL will think it's impossible. Because it's not a sure thing. Because it's difficult. Because it's a journey, not a one-and-done effort."

"What's the other one?" Lilly gestures to my finisher jacket. "Ironman Boulder?"

"Yes. Talk about feeling impossible," I sigh, and tell them the story of Eve finding me just after I ran down that hill at the beginning of the marathon. They loved it, and immediately grasped the meaning of the story.

We end our conversation with the owners excited instead of scared. In fact, their positive energy was so strong, that I asked them if they could encourage me when I'm having a down day. They said absolutely, stop in for some caffeine and a pep talk.

It feels great to have helped them. But I have an uneasiness about this. Who am I to encourage two tenacious young women to go for it and dive into opening their restaurant? So I share my concerns with Eve.

"Joel, you didn't push them into anything. You showed them what was in the way of their work."

"Yeah, but what if it doesn't work out for them?"

"Do you honestly think this coaching session will make or break the success of their restaurant? You're not that arrogant, are you, Joel?" Eve smiles.

"Nope." I sigh. I knew better. "My ego got in the way."

"It sure did." Eve and I walk to a nearby table. "But that's okay, Joel. You are practicing."

"Some practice," I huff. "I'm exhausted."

"I realize you're out of your box working with Lilly and Jessica. As much as I like you stretching yourself out of your box, the reason I have you work with them is to get hands-on practice. You are practicing *learning*, Joel."

"Doesn't feel like it."

"Too many people learning agile and lean think they can go to training, read a book, and be good to go. They think they can tackle the unlikely without practice. Because they don't understand that agile and lean are about changing the way you think."

"No silver bullet."

"Trite, but it's true. And of course, there are a lot of enterprising coaching and consulting companies out there. They are looking to sell a silver bullet to companies like yours, and leaders like you. Most of them are well aware of what they are doing. They know a company transformation is unlikely work. They know you're in a big hurry, and don't want to practice. You want the answer. You want to check the box and move on with your lives."

"Great, now I have that to consider, too," I sigh. "This is so very unlikely to work."

"Yes, Joel. They plan on long, profitable engagement of getting you going on the unlikely work. Parasitic help, because the company doesn't want to take the time to learn. Only achieve. Get the A at all costs."

Eve presses on, "The kind of practice I ask you to do at J&L's Café helps you *become* an agile leader over time, Joel. The kind of practice that, when Meryll sees you standing in a chalk Ohno's circle in the café, she is thankful it's you doing it, and not her. Because most people don't want to do this type of work, Joel. It's very hard, very uncomfortable, and it might not work."

"Practice."

"As an agile leader who practices learning, you won't have all the answers out of the gate, and you'll get comfortable feeling that, and saying that. You're going to learn what messages resonate with leaders and teams, and which ones don't. You'll learn what kinds of coaches work for your teams, and which ones don't. It's likely more than one kind. You'll even learn which parts of the company are the most challenging to change, if you haven't already."

"I'm beginning to guess which ones."

"You won't fall to pieces when something doesn't go well. And that will happen often. You will be able to help people recover faster and stronger from their missteps as well. Your team and other stakeholders will learn this from you. Maybe even Rick, in some small way. When you approach this work as practice, instead of as a final decision lasting for the rest of your life, you will become an agile leader."

"Wow."

We both pause for a moment.

"You didn't even get a chance to tell me about your race. Perhaps next session, if you're up to it."

"You're killing me. Again."

"I want to hear more about it. Especially about this Tony guy," She smiles. "But what I saw that day, and what you shared with Lilly and Jessica is really all I need to know about your race. You are really doing well at this, Joel. Because you learned how to work through the unlikely."

"Thanks," I sigh. "But the unlikely part of me leading at WL still needs your coaching."

Eve pauses, and patiently looks at me. She doesn't think I need coaching on this right now.

"Uh…" I'm missing it. What is it? Learning…I'm practicing learning…

"You got this, Joel."

Oh. Learning.

"Uh, if I'm supposed to be thinking what question I should be asking right now, and when I arrived here, I didn't know what my question was supposed to be. Now I think I know my question: How do I lead learning?"

Eve nods, "Yes. Ask yourself that, and you will have your next, right move. And, we are going to circle back on refreshing your vision. I think that fits in nicely with the first question. You don't have to do anything with your vision, except reflect on it, so we can adjust it to make sense for the future."

I sigh, "All righty then."

I make a note of these assignments in my notebook.

We leave our table and head for the front door. That's when we are stopped by The Mouth walking in. He looks from me to Eve, and lingers on her before punching me in the shoulder. Man, this is twice now at J&L's. Why is he here?

"Joel! How the hell are you? And who are you with, here?"

I greet Tony, and introduce him to Eve as a coworker. I just couldn't bear to tell him she was my coach. I'm not ashamed of having a leader coach, an agile coach. I just can't deal with Tony. The less he knows about me and my personal life, the better. He'd have a bunch of annoying questions and on it would go. Fortunately, Eve played along with the introduction. The look on her face might have looked neutral to someone who didn't know her, but I could see how amused she was with meeting the now infamous Tony.

"I think I know you from somewhere," The Mouth points at Eve, who gives him a faint smile. What can I do to get out of here?

Eve ducks out before either of us can say good bye to her, "Well, I will let you two figure that out. I have to head back to the office."

"I have to run as well," I pivot toward the door. "Good to see you." I quickly exit. I don't care if it looks like I'm chasing after Eve, even though she's headed in the other direction. I will do anything to escape.

Tony calls after me, "Let's get together soon, Joel!"

Reflections

Practicing learning is an active process that is uncomfortable.

Training, books and tools can help me practice, but they do not replace it.

Becoming an agile leader takes practice and time.

WL needs leaders who can help the organization practice learning.

Company transformation is unlikely work.

Transformation silver bullets don't exist, but are sold to companies like WL every day.

Certification Justification

I'm in a monthly leader meeting with all of Rick's direct reports. Well, except for Thad.

Rick introduces the first topic, agile training. Everyone must attend Essence of Scrum. Part of our transformation strategy at WL is for everyone involved with agile teams to get trained. He is taking the class next week, and invited our group to join him. I'm pretty stoked that Rick is acting like the sponsor we asked him to be. Well, this isn't exactly sponsorship, but his influence is needed for this kind of requirement.

"How long is the training?" Vladimir asks.

"Joel, here, can fill you in on the details after the meeting," Rick motions toward me. "It is a two-day class. Quite pricey, too. I know you're all thinking strategic expense management, and that is good. Training to change how we work and lead, for the better, can be justified. I believe in giving this transformation our best. If we don't get the training, how will we know if we are doing it right? I hope you all will get the training, too."

"Two days? That's a big commitment," Vlad turns to me.

Here it begins: the practice of inviting people to change. Inviting them to learn. So practice I will.

I have to be careful here. I am not asking people to go to training for me; I'm asking them to go for themselves, their teams, and their company. It won't please me if everyone goes to the training; we'll just be better for it. I'm comforted that Ben, the executive sponsor of the very first agile team at WL, is in the room. He has also already attended Essence of Scrum.

"Is it really the expectation that *everyone* receive training?" Vlad asks. "Even the teams? That seems like a lot of people, and a huge expense. We need to be careful not to fall prey to greedy training companies who don't really care about our success."

Ben clears his throat, "Vlad, that is a really good point about being vigilant. Joel and I are on the same page with that, and will challenge this team and others to keep that mindset. However, training is a necessity to success. The teams who are going to practice scrum attend a five-day class; the rest of us get to attend a two-day session."

I'm not sure I would have said that to Vlad right now, but I'm not going to talk over Ben. He may be able to help Vlad more than I can. I'm just the transformation leader; Ben has actual scrum teams in his area, and they are peers.

"I'm not against changing, really," Vlad assures us. "I just don't believe a two-day training class is the best use of my time."

At least Vlad is thinking about value. Ben looks to me. My turn.

I tell Vlad that there is great value in leaders like him attending training, if we are to support and lead the change. There are many changes that need to happen in WL beyond technology; our actual environment needs to change if we are to survive the disruptions happening in our industry. Things like how we fund projects, how we reward people and teams, and even how we grow leaders need to change. These are called barriers to change, and teams can't overcome them on their own. We're going to be asked to help them change the way they work, so we need to understand them. In fact, there is a small group of business partners who are also going to attend. Some of them are the ones who attended our first transformation report outs; the others were brought along by these people.

Vlad's eyebrows furrow, "Isn't that why we have C²? Culture change?"

"That's something different, Vlad." Meryll tries to help.

"What's C²?" Ben asks.

Gabriel looks at Ben, "You haven't heard?"

Ben shakes his head no.

Gabe huffs, "Then there is that fill-out-the-spreadsheet-about-your-team's-skills-in-24-hours crap."

"Well, how did the business partners get involved?" Vladimir is concerned. "What if they go to training, and then go off on their own? They could take us seriously off track."

"That could happen to tech people, too, Vlad," Meryll assures, and then turns to Gabe. "*Spreadsheets*?"

"That's true," Vladimir is thoughtful, and oblivious to the other noise at the table.

Gabriel gets Ben's attention again, "Ben, C², it's a culture committee."

Ben sits back in his chair, "Huh. A culture committee. At the same company where we have a scrum team that gets yelled at for putting flip chart paper on the walls."

Environment, culture, scrum training, oh my! This is feeling rich. How deep do I go with the group? I feel so responsible for their adoption, and I shouldn't. I will have to talk with Eve about that.

"Vlad, C² is a cross-functional committee organized to change the culture of WL. Both Gabe and I were asked to be on the committee. Essence of Scrum training is specific training for those who will lead or work with scrum teams. They are very different activities, but they both contribute to the overall transformation of WL."

"And the spreadsheets with skills?" Meryll asks.

"Black box," Gabe sighs. "We should discuss within this team."

Just then, Thad walks in, and sits down next to me. At least this way, I don't have to look at him.

Rick greets him, "Thad, we were just discussing how I'd like everyone from this team to attend Essence of Scrum."

Thad opens his laptop, "That seems like a really good idea. How long is it?"

Great, now we are revisiting this garbage, just because Thad was late. Actually, we are repeating it all because Rick took control of the meeting and steered us that way. He doesn't want Thad to miss a thing. Golden Boy.

When Thad learns the class is two days long, he nods his head, "Good. If it were only an hour or two, I would say we need to find better training. Of course I will attend."

Rick thanks Thad, and Vladimir looks expectantly at me. I guess he needs more purpose to agree to this training commitment.

"Vlad, I know it's a big commitment, but I believe there is value in this training and I've shared what that is. It's worth a try. There's also a test you can take after the training to get certified in the basics of scrum."

"Certification?" Meryll asks.

"Just spitballing, here," Thad sighs at his laptop. "How many certified leaders does WL actually need in order to reach our goal?"

"Well, I don't think you need a spit ball to see that this could easily spin out of control," Vlad gently warns. "I hate to be the one to always think of the challenges, but our organization has a way about jumping head first into things without a full analysis. We go after the shiny object, and lose all sense of purpose."

"And then we have to concern ourselves with the paradox of whether to train the contingent labor force on scrum, or agile, for that matter," Thad looks at the ceiling.

"We trained the contractors on the scrum teams," Ben shrugs. "They are, after all, doing the work."

"Did the contractors pay for the training themselves?" Vlad asks.

Jack's eyes catch mine, and show his amusement with the ping-pong session unfolding before us.

Rick waves a hand like he is brushing a fly to the side, "Anyone on *this team* can get the certification. I want us to be strong in our understanding of scrum, and I want us to lead by example. As for a full analysis, the recent events with our competition and our regulatory fines are the only motivation I need to change. Whether there is a plan, a TPOC, or any other nonsense of the like, in the very near future, we will discuss the training needs for the rest of the organization."

There is a short pause. It feels longer, because the discussion we just had was exhausting. Well, at least it was to me.

Vlad nods, "Fair enough. I will attend the training, and then encourage all of my teams to do the same."

"We really should hold off on everyone getting certified," Thad says. "I'm being transparent, here, I think we are putting ourselves at risk if everyone has this…this certification."

"Why is that, Thad?" Now it's Ben's turn.

"Enough." Rick shuts it down. "We will defer to Joel's transformation team on who should get certified, and who, if anyone, shouldn't."

"What if you train this entire organization, and they leave?" Thad looks at the ceiling. "I can see it: Sheepskin in hand, they have a shiny certification that is attractive to other employers. We'll be left high and dry."

Well, I *thought* Rick shut it down.

Jack clears his throat, "What if everyone stays?"

"What might *that* mean?" Thad's eyes actually wander around the room to find who was talking.

"We can't know what it means, because it hasn't happened yet," Jack shrugs. What a lame deflection.

"That's true," Thad returns to his laptop.

But it worked.

With visible irritation, Rick shuts down the conversation again. Get to training, when we meet next month, everyone bring their certification to show to the group. It's the only way we currently have to measure the steps we are taking toward becoming an agile organization. We will *ooo* and *ahhh* over our accomplishments, and then roll up our sleeves to support our teams. Rick's message was solid. Some days, he sounds like he really gets it, and I'm encouraged. I'm glad today is one of those days.

"Well, that was fun," Meryll sighs as she sits in the guest chair next to Jack, who is texting and chuckling to himself. Normally, Jack doesn't have his phone out during our discussions. She leans over toward his phone, and he pulls away.

"You're texting Anika?" She looks at me. Anika used to work at WL, until she could no longer take the toxic leadership behaviors around her. She's been gone for a few months.

"Nah." Jack is still laughing. "I have to tell you and the Sultan, here, a great story."

"Oh. My." Meryll giggles. "You two are dating."

This woman doesn't miss a thing.

"Maybe. But that's not the story."

"Then what?" Meryll demands. This short-term suspense is killing her.

"Meryll, maybe she's recruiting him to her new company. Since Jack's going to get certified in scrum, he will be a hot commodity."

"Right, Joel," Meryll rolls her eyes, and then turns to Jack. "You *wanted* me to see, didn't you? How long has this been going on?"

"Do you want to hear the story I heard, or not?"

"Lay it on us," I say. "After our last meeting, we are ready for anything."

"Good. You need to hear this one."

Apparently, Anika was attending her niece's fourth birthday party. It was a big bash with lots of family and kids. There were even pony rides out on the street for the kids. Later, there was a clown. Rockin' Willy the Clown. The kids were enjoying his show outside on the large patio, and Anika watched from a distance, along with a few other adults.

Rockin' Willy wasn't dressed like a traditional clown, but he was definitely a clown. The style wasn't pantomime or trendy or French; it was western. Artsy, upscale, western. No polka-dotted one-piece costumes, no oversized shoes, no horn. He had a velvet, three-piece suit that had a cowboy or western flair to it. It had huge pockets, which made it sort of clown-y. And a velvet cowboy hat. It had to be custom made.

Rockin' Willy's act was musical comedy. He had an acoustic and electric guitar, and pretended to be really bad at playing it and singing. He used songs the kids knew well, and then either sang badly, got the words wrong, or struggled to properly play the guitar. His act was polished, simple and happy. The kids loved him; even the parents were laughing. Some of the parents were talking about getting Rockin' Willy for their kid's party. Someone else said that he's very difficult to schedule because he's so popular.

Anika thought there was something familiar about him. Even with all the makeup and the costume. He was really tall, and had a deep voice, and intense, blue eyes. She passed it off at first. There is no way anyone she knew would put on a Rockin' Willy the Clown act. Yet the more Rockin' Willy entertained the kids and the parents, the more she realized she knew this guy.

Meryll puts up both of her hands at shoulder level, "Stop. Just stop."

Jack pauses shaking with laughter.

"Come on Meryll, you know you gotta hear Jack say it."

"No, I don't, Joel."

I start laughing, too.

"How long have you known about this?" Meryll asks.

"Just this weekend. The text I was reading when I arrived here—"

"Stop!"

"…Said *be sure to say hi to Rockin' Willy for me.*" Jack falls into another fit of laughter.

Meryll is beside herself.

I put my head in my hands, "Ugh. Rockin' Willy the Clown."

"Don't you want to know what our VP did next?" Jack asks. "It's so savage."

"I can't." Meryll shakes her head no. "I can't take anymore."

"Then you better leave, Meryll. Because Jack has the story of the year, and I'm not going to miss any of it."

Meryll slides down in her chair in surrender.

Jack looks at me, "I take it she wants to hear the rest."

"Please go on, Jack. Please tell us what else Rockin' Willy did."

After the acoustic act was over, Rockin' Willy the clown talked a little with the kids and parents. He handed out a few business cards. They all sang happy birthday to the four-year-old, and then it was time for Rockin' Willy to leave.

As Rockin' Willy packed his guitars and his gear, Anika headed to the kitchen, and remained there alone. She thought this would be the best place to avoid being seen by him on his way out. She can hold her own. No need to invite Thad drama back into her post-WL life. Anika made sure she had her back to the doorway, just in case Rockin' Willy wandered in the kitchen. She got out a cutting board, and worked on peeling and slicing apples.

Well, it didn't take long for Rockin' Willy to wander into the kitchen and head straight for the fridge. Anika was doing her best to focus on her task.

"Then how did she know it was him?" Meryll asks.

"He was softly singing one of his songs. *She knew.* So she became very busy making deals with God that she would never, ever sin again."

"God. I can't stand it." Meryll rubs her temples. "It's so…weird."

"What did he take from the fridge?" I'm pretty sure it was beer, but I had to know.

"Beer." Jack's shaking with laughter again. It's contagious, so I start, too. "When she heard the bottles clink, she stole a full glance at what was happening, and then went back to apples and making deals with God."

"How many?"

"Two longnecks. Dropped one in each of his big clown pockets, and then grabbed a third to carry out." Jack rocks his head from side to side.

"No!" Meryll reaches over in a Seinfeld manner, and forcefully pushes Jack's shoulder. He doesn't care. He's still laughing too hard.

"What else?" I ask. "You haven't told us the best part yet, have you?"

Jack slowly shakes his head no.

Meryll closes her eyes and sighs long.

I'm laughing in anticipation of what on earth Rockin' Willy the clown could have done next. My eyes are watering and my stomach hurts.

Rockin' Willy was a hungry clown, so he made himself a few ham sandwiches, and stuffed those in his big clown pockets, too. Then, the humming stopped.

Anika held her post. She sensed he was quietly walking near her. The whole scene was already too much to take. Slice, slice… and then suddenly, SMACK! Rockin' Willy slapped Anika's rear end while letting go a much-practiced clown laugh.

Even though Rockin' Willy tried to run away, her instincts immediately took charge: her leg flew out to kick her assailant. She made direct contact, and sent the clown airborne. He landed flat on his back with a huge thud. She could literally hear the wind knocked out of him.

She wasn't finished.

All her fears of Thad drama aside, she placed her Alice + Olivia pump firmly on Rockin' Willy's crotch and said, "A western freak like you should know you don't mess with Texas. You touch me again, or touch anyone else here, you'll be eating your balls."

Their eyes locked for one second. He looked concerned of what would happen next, but it was clear he had no idea who she was. Still, Anika wondered if he was just a good actor.

"Get your pathetic, purple ass out of here." She roughly released her foot from Rockin' Willy's nether region, causing him to groan. She walked over to the side door of the house. Arms folded, she stood and waited for him to pull himself up and head for the door.

"And gimme the longnecks," she motioned at his full pockets.

Quickly, he fished them out and handed them to her. He was gone in a flash.

My office is silent, except for the three of us gasping for air between fits of laughter. What else is there to do?

"Anika thought she might faint," Jack sighs. "She still wonders if Rockin' Willy knew who she was that day. I had to talk her off that ledge. I reminded her that there is no way in hell Mr. VP of lean wants anyone to know he moonlights as Rockin' Willy the musical clown."

"Rockin' Willy the musical, beer-stealing, sandwich-stealing, perv clown," Meryll wipes tears from her cheeks.

"Exactly right," I say. "Jack, that has to be the best story I have heard about anyone at WL. Ever."

Meryll perks up, "It is! I can't think of anything remotely close to this one."

"So, there you have it, Joel," Jack holds both of his palms up. "A new perspective to appreciate about our VP. And I have a new perspective to appreciate my brave, hot girlfriend."

"Yeah, Rockin' Willy the Clown is a cowboy in so many ways."

"I don't get it," Meryll says.

"Meryll, Thad goes off on his own at WL. He's a cowboy, a lone operator."

"Got it. Just what we need," Meryll sighs. "What are you going to do about him?"

"I'm not sure I'm supposed to do anything about him. I know I need to change into an agile leader, but I don't think that means I have to do anything special for Thad."

"Well, at the very least, I think he needs to attend sensitivity training."

Meryll laughs and pushes Jack's arm, "You should ask Anika to call you Rockin' Willy and—"

Jack cuts her off, "Shut. Up."

"Well, this is more serious than I imagined, Jack," Meryll is impressed. And like me, happy our good friend has a worthy companion.

"She loves her new job. She's got all sorts of opportunities that she never had at WL."

"Of course she does."

Reflections

I feel so responsible for my peers' adoption, but I shouldn't.

Essence of Scrum training might be a bigger issue than I thought.

Certifications complicate learning.

Cowboys cause confusion instead of adoption.

RIE Report Out

I'm attending the report out of the first RIE at WL. That's a lot of acronyms, but I'm going to roll with it. I didn't receive an email invite, but Vlad made sure I got the event on my calendar. I wasn't too concerned about crashing, but no one seemed to mind that I was in the room; they were too busy preparing for the event. Guests like me were stakeholders who had never experienced anything like this, so they weren't counting who was who, either.

I didn't see anyone from HR or corporate planning in the room, but I could have missed them. Maybe that's intentional. I could imagine how those organizing this first event might not want the corporate weight of HR and funding upon them. Yet this first event is also the perfect time for them to get involved. Don't delay the difficult conversations; have them now at the beginning. Expose the problems and find a way to move forward.

The good news is that we did have business partners there. Some were guests, and some were part of the RIE itself. I'm relieved and encouraged to see them. As much as Thad is not sharing what he's doing, he is mostly involving the right people.

We were all stuffed into a very small room. The walls were covered with butcher paper, sticky notes and red dots. All I could think of was facilities. Even with the painter's tape to care for the walls, the facilities folks would have a conniption if they saw this room. That reminds me, I should circle back around with Lucy from facilities. She was supposed to join our transformation team, but I haven't heard from her in a while.

The RIE was for an operations group's process improvement. What my company does in this specific area isn't important; the key is that it's a repeatable process that was determined to need improvement. The RIE helped the team of stakeholders, leaders and practitioners have time to really focus on one, specific process. They've been preparing for it for over a month. People were trained, and stakeholdering happened. Mostly.

Over a month? But…isn't this part of our transformation? Thad is clearly involved in this event. It's not aligned work, but speaking up about it now is not the right time.

There were two coaches in the room helping the team get ready, but they did not run the event. Were they there just for this event? Maybe they worked with them in the days leading up to this event. At any rate, the team ran it with the coaches standing by, supporting them. The presentations were a little shaky, as expected. This was their first one. And having Thad for a sponsor, leader, or whatever he was for them, is enough to make anyone nervous.

It's exciting to see this team and this process get attention so that it can improve. I work to step past all that is wrong with this event to take in the value of it.

Karen stands next to me. She notes the coaches as well. I ask her if she was privy to any of the hiring of these coaches, and of course she wasn't. We agree to let it go. There are so many problems that we could get hung up on with Thad operating independently of us, but not today. We can always address that later. Not much later, just not during this event. The people in this RIE deserve our attention and our support.

During the report out, I was reminded of the work our first scrum teams did. They both got training, some coaching, and then worked in an isolated environment to get started. It is hard to work this way, but it's possible. Just like the scrum teams, this group accomplished great things in just four days. They are energized, and ready to take their improvements forward.

The question and answer time at the end of the event was silent. A revolutionary way of working, and no one is questioning it? Or, are people fearful of questioning it? I do see smiles of approval…

I feel the need to say something, to ask something.

I sigh, just imagining the conversations I will have to engage in. "I'm up for the challenge."

Karen and I leave the event together because we are going to the first floor to see the progress of the new home for our three scrum teams. As we make our way through the main walkway for WL, we see some large signs being installed on the wall.

"Would you look at that, Joel," Karen stops to watch the work.

We take turns reading aloud the signs that are already on the walls, or are being installed.

"*C² & U*," I read.

"*Think Happy. Think Strong*," Karen sighs.

"*Work Happy. Work Strong*," I say, rather quiet. I'm getting deflated.

"*Be Happy. Be Strong*," Karen plays up her voice with feigned conviction.

Just then, Meryll walks up to us. We greet each other, and I am enjoying her incredulous face.

"What…is…this?"

"Exactly," Karen smiles. "I think it's supposed to motivate us. Are you feeling it?"

"Does everyone even know what C² is?" Meryll asks.

I laugh, "Does it matter?"

"*Be happy. Be strong*," Meryll recites. "I don't like being told to *be* something. Does someone actually think that by putting up big, fancy signs that we should listen? I feel like I'm being forced to change. And…I have no idea what part of my culture I should change."

Karen shrugs, "Yep. It's just another item our transformation team will have to discuss, so we can field questions about it."

"That's why we've got you on the team," I say. "Aside from being smart, and an organizational change wizard, you have a sixth sense for how this stuff is going to play out with real people, not the corporate communications people."

"Eventually, won't you have a voice with them to stop this kind of crap?" Meryll asks.

"We'd love world domination, but that's not really our goal," Karen is wistful. "I think all we can do for now is keep bringing them along, and hope that eventually, they begin to get it."

Reflections

People who are the pioneers for change have a difficult road.

Business partners are a strong part of our transformation.

When will we be an *us*, instead of tech/business partners?

Our company tries very hard to force change.

Scrum Master

"Hi Joel. Thanks for taking the time to meet with me."

"Of course, Waylon. What's going on?"

Waylon tells me he has a concern related to the agile transformation at WL. Specifically, his team, *Can't Make This Up*, and the leadership of the transformation. He discussed it with Ben, and the two of them determined the best approach is to talk with me.

Now he has my interest. I've been meeting with many people lately, getting the transformation vision and message out. Mostly, it's been a lot of very scared people coming to me, worried that they are going to lose their jobs because of the change. Given that Waylon is the first at WL to have an agile role, well, maybe the second person, I was fairly sure he wasn't worried about his job.

Two days ago, Thad appeared at the team stand up, a daily meeting held for just the scrum team. The purpose of the meeting is for each team member to talk about the work they are going to do today, and if they have any impediments or learning opportunities for other team members. The daily stand up is just for the team; leaders and others are not normally a part of it. All of this is taught in the Essence of Scrum class. Of course Thad hasn't attended the class yet, so he wouldn't know that.

Waylon wasn't concerned about Thad showing up to see his stand up; he was concerned over *why* Thad showed up at the stand up. He wants to learn how to be a scrum master.

Huh. Rockin Willy the Clown is a scrum master, too.

"Yeah, he said he was the new scrum master for his team, so he wanted to learn how to do the job. He didn't really ask me if I would teach him. He sort of...implied it, or announced it."

"Scrum master," I repeat, because I can't imagine it.

Waylon carried on with the daily stand up, while Thad took notes. When the meeting was over, Thad asked if Waylon had time to answer a few questions. Waylon told Thad he had 15 minutes before he had to attend a refinement meeting with the team. Thad suggested that he could be late to the meeting. Waylon drew upon his Essence of Scrum Master training, and pushed back at this leader who was going to get in the way of actual work getting done. Waylon said he couldn't be late for the meeting, but he'd be happy to meet with Thad after lunch. Thad said he was busy with an *important development*, and could only meet right now. Waylon apologized, and said he could talk only until the refinement meeting, which is now in ten minutes. At this point, Waylon was sure he would be fired, but he wanted to stick to what he learned. He was told in training there would be moments like this with leaders, but he didn't think it would happen with Thad.

At any rate, Thad said he understood, but then questioned Waylon on why his scrum master job made him so inflexible that he couldn't drop everything to meet with a VP. What to say? Waylon chose what he was trained on: his intent.

"I told him the entire team was getting together to talk about and prioritize the work we had in our backlog. I invited him to come to the meeting to observe and ask questions later, but he wasn't interested in our work. Joel, I really tried to convince him that the refinement meeting is a great way to see how a team works together with the product owner and manager, but he wasn't interested."

Waylon invited Thad to return the next day for 30 minutes, after the daily stand up, and Thad agreed. The next day, Thad showed up again for the daily stand up, instead of after it. He took pages of notes in the 15-minute meeting.

When they met, Thad began asking all sorts of questions about scrum and agile. Real basic things that are covered in the Essence of Scrum training, such as what does the team do all day? How does the team know what work to do? How do you measure progress? On it went. Waylon answered the questions, and each time talked about the Essence of Scrum class offering great help with the concepts.

"What do I do with this guy? I don't have time to train him on scrum, but I don't want to be a jerk about it. When I talked to Ben, he said we should be real careful. I told Ben that we are the first team, we set the precedent. If we're going to let leaders break all the rules right out of the gate, we'll never make it."

I commended Waylon on his patience with our curious leader, and for drawing upon his training to help drive his behavior with Thad. It's true we need to be cautious with Thad and leaders like him, but getting a healthy start is part of our vision for the transformation. I have more concerns with Thad's behavior that I don't share with Waylon, but I will definitely need to take action on them soon.

"Waylon, I'm not sure, but it seems to me Thad is after something more than just understanding your role as scrum master. I really don't know, but that's how it feels."

"I think he's assessing us, Joel."

"For what?"

Waylon shrugs, "Who knows. We're a new team, I have a new role. He probably wants to make sure we are productive and not squandering money. The problem is, we are just beginning, and aren't saving any money. In fact, we may never actually save money. I'm concerned he or someone he works with will shut us down."

Man. Waylon's right. Thad is on a hunt for metrics.

"The other thing is, he is serious about the scrum master role. He thinks he is one, and that he has a scrum team. I asked him who his product owner was, and he said it was him. I didn't have the heart to tell him that was impossible," Waylon sighs.

I thank Waylon for making time to talk with me. I offer him encouragement for going first. There are benefits to going first, and then there are challenges. I encourage him to keep sharing that message with his team, so that when the entire team feels pressure of one kind or another, they won't crumble under it.

"Yeah, we are practicing how to talk about what we do and why we do it this way, without being jerks. So far, one of the most difficult things we've encountered is explaining what we do to teams who want to be like us."

"You don't want to make them feel left out."

"They already feel that. We don't want them to feel like we are right and they are wrong. We are just the team going first. I think some people absolutely get it, and are even glad they're not the ones going first. But other people are offended that their team wasn't the first to be selected. Now they resent us. Some of them will probably go scrum on their own, because they can't wait to be asked."

"Really?"

"Yeah. It's not a bad thing in theory, but we have to organize our work the right way. If people start going off on their own, this can stress all of us. I'm getting off track, there. The bottom line is, I'm not sure what else to do about Thad. Well, I know what I want to do about him, but I'm not sure that's acceptable."

"Waylon, I trust your judgement. And, I have your back, so I support you in wanting to do the right thing for you and your team."

"Thanks, but I just didn't want to shut this guy down on my own."

"You can. I'm not going to tell you what is and is not acceptable in this case, because you are trained in it."

"Certified, even."

"Right, and Thad is not. But he can be, and so instead of getting one-off training from you, he should be encouraged to attend Essence of Scrum."

"I will tell him."

"Great. I support your decision and agree with it."

I am still concerned about what else Thad is looking for regarding metrics. Hopefully, this curiosity will be reshaped by attending Essence of Scrum. I need to reflect on this one. Do I take the proactive stance and approach Thad with what I've heard, or do I just let it play out? I encourage Waylon to talk about this further with Ben, and if needed, the three of us.

Waylon looks at me and smiles, "Dude, you know your stuff. Did you get some kind of agile leader training?"

"I have a coach," I smile back. "I have grown a lot in the process, but I still have so much to learn. My coach pushes me to learn to be a better leader, and I practice daily."

"We are getting an agile team coach in two weeks, thanks to Karen and your team. We really need one."

"I can't wait to hear how your team benefits from one. Having a team coach with you all day sounds like a great way to get a healthy start."

"I hope so," Waylon pauses. "You sure you're okay with me suggesting to Thad to get training? I mean, if he approaches me again with a bunch of questions."

"Absolutely."

"Thanks. It's great to have your support."

"It's the least I can do, Waylon."

Reflections

The scrum master role is not a job that just anyone can do.

Thad is using positional power to force learning, which can't be sustained.

Metrics mania is about to begin!

Tourists and Cowboys

Meryll and I are out for an easy run together. I've had a long rest after Ironman Boulder; it's time to get moving again. It had been a long time since we ran together, so Meryll was pushing me to join her. I was hesitant because I'm sure my legs are still demolished, but Meryll promised she'd go easy on me. I had nothing left to do but agree, and then ask Cele or Caroline if they could watch Meryll's kids while we run. Caroline was working, but Cele and the boys were happy to help. Never to be left out, six-year-old Cici claimed she was going to help, too.

Ever since Meryll began agile leader coaching like me, she has made radical, lasting changes in her life, and her career. She claims she probably would have still changed, but not for years down the road, and probably not as radical of changes. Coaching opened so many parts of her mind and heart, she transformed her life. She divorced her husband, but only after he had gambled away everything the two of them owned. Then, she began working on taking better care of herself. She stopped drinking giant, sugared-up coffee drinks, and picked up running again.

I've seen these great outward changes, and then I've seen her change on the inside. She's more positive, smiles more, and laughs more. And professionally, she is working on being a better leader. Just like me, she solicited her team for feedback on her leadership style and impact. The team's feedback was brutal, so bad that she was convinced she would get fired, and that it was the end of the world. I helped her pull out of that dark place by simply reminding her the coaching we are doing is separate from the performance management process at WL. In fact, it's in our contract with the coach. I also reminded her that my feedback was about the same, and I lived to tell about it.

So now we are in better places, enjoying a run together. The last time we ran together, I pushed Meryll to consider registering for an Ironman. She didn't say no, so I'm going to keep on that. I think she'd really enjoy the challenge and the accomplishment.

"What's going on with the transformation, Joel? What's the juice?"

"You get all the juice, Meryll. I don't hold back anything."

"I'm anxious to help you. What can I do to either help your transformation team or help you?"

I laugh, "Careful what you wish for, Meryll."

"Seriously. I have my own work. But I believe there is a way to help you while doing my work, and new work."

"I'm sure there is." She's looking for me to give her an exact assignment, but I'm not going to do it.

"So, how do you think I can help?"

"Meryll, I'm not being a jerk when I say this: I want you to discover on your own what you should do. I can involve you in the big ceremonies and activities. From there, you will be able to see where to connect your work to what's changing. And, keep attending the sprint reviews for teams like *Can't Make This Up.* They keep bumping into some complex barriers that might require a team of leaders to push through."

"I get it. I knew you wouldn't just tell me what I could do." Meryll laughs.

"There's a ton to do, but we must make sure we are collaborating in a coordinated way. Attending the transformation team ceremonies and the sprint reviews of our first scrum teams will set us up for success."

Meryll agreed. Having Rockin' Willy the Clown running around on his own is enough; we don't need a whole bunch of people doing their own thing. She promised to do her best to connect to the transformation work.

"I also want you to promise me to not be a rule follower."

"I don't get it."

"It's the opposite of cowboy."

"Oh. You know, I just read a blog about this. It was called *Tourists and Cowboys*."

"Huh."

"And, I can use this label, because I grew up with it."

"What do you mean, Meryll? You were driving cattle in a prior life?"

Meryll laughs, "I am the daughter of an honest, actual cowboy. My father was varsity Rodeo at University of Arizona, and elected Rodeo Boss three years in a row."

"Well!"

"Yeah, I grew up watching Marshal Matt Dillon dispensing justice and righting wrongs. From what I can see, cowboys get a bad rap in the agile world. It's the sheep herder versus cattleman range war all over again."

"I will be careful with how I use it, Meryll."

"I don't think you need a trigger warning, Joel."

"Trigger was the beloved horse of singing cowboy Roy Rogers."

Meryll laughs. "Very good."

"Tell me about the agile cowboys in the article. Not of the western clown variety."

"Well, tourists are the people in an organization who adopt the change, but not until someone spoon feeds it to them. They want you to show them what to do, and how to do it. It's passive. And safer that way. They attend the meetings, smile and show good support, but then after the meeting, they ask what they should do next." After all, if it's not in their performance expectations, why change?

"Ah, I get it. Another kind of tourist is the one who sits through my 60 or 90-minute presentation on why we are changing, what we are changing, and then after the meeting, they ask me for talking points. They are disappointed when I tell them there aren't any."

Meryll laughs because she can't believe people at WL have asked for talking points. Then she said she should know better; most people at WL are not going to carry things forward to their organization unless it's perfectly lined up.

"Joel, I think Thad is one kind of cowboy. His intentions are not that great. He's got an agenda, and a secret plan. The blog didn't even touch on this kind of evil cowboy."

"Great, we have the anomaly."

Meryll describes the other kind of cowboy as a person who has good intentions, but doesn't realize they are taking energy away from the focus of the transformation. They are full of enthusiasm and energy, and likely fresh out of training. They can't wait to get started, so they go off on their own and often begin making up their own rules. This is how small problems turn into messes that can't be ignored. Then, the transformation team and others have to shift their energies to sort out, fix or mitigate the mess.

"Well, sending our leader team to training, we are right on the cusp of this one. I'm so glad we are talking about this now. I need to talk to Rick about it so he can help lead our group. We gotta use GEKO."

"GEKO."

"It's an acronym for creating a sustainable change. The G stands for get a healthy start. Something I'm really trying to take to heart. We won't be perfect, but there are ways to begin that help us better sustain the change overall. Leading with the vision we created is part of that."

"Well what about the other letters, Joel?"

"Uh give me a second," I think I can remember all the parts of the acronym. "Exercise to see the work, kick off right, and then optimize through plan, do, check, adjust. *PDCA*. Ha, there's another acronym."

"Well, we are doing the exercise part right now, but I'm sure you mean some kind of discussion with the people who do the work. Is that the same as an RIE?"

"No. Well, something could come out of it that helps us see the work. But an RIE is a narrowly focused process improvement opportunity. The exercise I'm talking about is very focused on only seeing the work, the flow of value. Not about how to improve it. That will happen later. But with the way Thad is running around spinning up RIEs and a team to lead them, this change is going to be very difficult."

Neither of us really noticed we were running until Meryll took a look at her GPS. We've already gone four miles. She's impressed with how well I have recovered from my Ironman. I guess I am too. More impressive, I never would have imagined that we could talk about changing WL for four miles.

"Well, here's some good news: The blog said people need a filter for the change. And that the filter is the vision, so you are right on track, Joel!"

"That's something you can help with, Meryll. Share the vision for why we are changing. Relentlessly, but connected to the work."

"I can do that, Joel. And you should probably do more than one-off meetings with Rick about the transformation vision. You know, get him formally on track to connect people to vision."

"Excellent point. I have dropped the ball on that."

"Joel, you can't do it all. I'm happy to be a transformation whisperer," Meryll laughs.

Meryll has a great point, so I tell her that. We talk about how I can't put this transformation all on me. In fact, I have a reflection in my notebook about that. I'm not responsible for people to change. I'm responsible to illuminate the change and connect people to it. What they do with it is up to them. Oh, and I guess I'm to help build some guardrails for the tourists and cowboys. Meryll liked the guardrails concept. No prescriptions or recipes for change, but we do need to keep people on the path.

We're nearing the end of our run, and Meryll is again impressed with how well I've recovered. I remind her that I have had a huge break, so I'm well rested. The real challenge will probably be a few weeks from now, when I get back into a regular training schedule. Now feels like a good time to encourage Meryll to register for an Ironman, so I do.

"Come on, Joel. You can't be serious," Meryll laughs. Nervous laughter. The kind that tells me she's been thinking about it.

"You've changed your life so much for the better, Meryll. This challenge is perfect for you. And the accomplishment feels incredible. It still does today. And many other days. I'll be sitting in a long meeting, or stuck in traffic on the way to work, and I think back to that day and how great it felt to finish."

"I don't know, Joel. I've been thinking about it since you brought it up. Some days, I think it would be a great challenge for me. Then other days I think of the kids' activities, the amount of training needed, and it seems just about impossible."

Impossible. There is that word again.

"What about it seems impossible. I'm not being flippant either."

"Well, it's just…the distance…I think it comes down to training to be able to bike 112 miles. And, what if it's another blazing hot day like you had? I can't believe you pulled your tail for 140.4 miles in that heat, Joel."

"140.6 miles."

Meryll laughs at herself, "See? I don't even have the distance right."

I zone in on her statement about biking 112 miles. What about it seems impossible: the actual 112-mile bike ride in the race, or *training* to ride 112 miles in the race? She tells me both.

"It seems impossible, Meryll, but really, it's just unlikely."

"Whatever. Impossible, unlikely. I just don't think I can do it."

"That's what makes it impossible. You won't try. If you register for it, then it will only be unlikely that you can do it."

"Well, aren't you philosophical today, Joel! I can see there is no getting out of this conversation, is there?"

"I will register too."

"You'll do another one," Meryll says flatly. "After you just finished on one of the most outrageous days ever."

"Yep. We can ride together. Or, even better, I can hook you up with Tony."

"There will be no hook ups, Joel."

"Maybe we should rope Jack into this. You know, shame him into coming to the dark side."

Meryll laughs, "Yeah, he should join us. He can impress Anika, or rope her into doing it, too."

"Come on, Meryll. You know you want to do it. You want to see just what you're capable of. Which I already know is anything and everything."

"Fine. I'll register for next year's race. Damn it, Joel, you are killing me."

I can't believe she said yes. I thought it would take a few years. This victory is a direct result of Eve's *unlikely* coaching.

"Sweet! Let's register today. Before we change our minds."

"Right. You want me locked in and loaded before I realize this goal is totally impossible."

"Nope. If you register, anything is possible. And certainly, it's unlikely that you'll do the training and get to race day healthy, and then finish the race."

"Wait a second, Joel! You just said I could do this."

"Hear me out, Meryll." I say as we are nearly back to my house. "This goal is very hard, incredibly daunting, and yet, so very satisfying. The reason it's satisfying goes beyond a finisher medal and shirt. It's about the unlikely-ness of the challenge. I want you to experience the satisfaction of doing something so unlikely that most people never even attempt it. It's one of the best feelings, Meryll."

We're back at my house, stretching. When they hear us talking, the kids migrate from the back yard to the front.

"That's the best explanation I've had from anyone about why they do something outrageous like an Ironman or the Leadville 100 race." Meryll smiles.

I shrug, "Let's do the unlikely together, Meryll. For Ironman Boulder, and for the transformation at WL."

We high five, excited about the possibility of accomplishing the unlikely.

Reflections

Cowboys can cause problems that hold back change.

Most cowboys mean well, so if someone spends time with them on vision, they can get on track.

Tourists wait for perfect conditions.

Tourists will agree to anything, but then not follow through unless there are guaranteed success.

A shared vision makes a clear path for everyone.

Relentlessly sharing vision makes it challenging to remain a cowboy or tourist.

Helping people see themselves accomplishing the unlikely feels great.

Protecting the Underperformer

The kids and I just finished watching the Broncos lose when Caroline asks if we can talk about work. We chat in the kitchen, while the other kids scatter outside the house to play football. I love how little Cici jumps right into the game. Everything is possible to her. Even playing in a pickup football game with her brothers.

"I wanted to ask you what you thought of this whole Rhoda situation."

"Oh that's right. Rhoda is the one who you think gets special treatment?"

"It *is* special treatment, Dad. I know you get this stuff, and like, I wanted to hear your opinion."

"What's the rub with you? What's the problem?"

Apparently, the entire water park team has required weekend days they must work, including holidays. Rhoda doesn't have the same requirements. No one knows why. Everyone speculates that she can do whatever she wants because she has been there so long.

"How old is Rhoda?"

"She's like, 153."

"Seriously, Caroline. Is she your age, college age, my age…"

"She's probably older than you, Dad."

I wince, "Ooo, and she's a lifeguard. Huh."

Carline rolls her eyes. "I'm not like, prejudiced because she's old. But she probably couldn't pull a five-year-old out of the water, Dad. She just like, stands around and does nothing."

"Okay, so there is a lot going on here, but basically, she is treated differently than everyone else on the team, is that right?"

"Yes."

"Is there anyone else who gets special treatment similar to hers? Or is she the only one?"

"Rhoda is the only one. Well, except for the lead guards. They have exceptions, too, but they just sit around in the office and waste time. At least they have to show up for their required weekend and holiday shifts. And you know what else about Rhoda? She doesn't participate in all of the guard training we have to do."

"Is Rhoda good friends with anyone there, even a leader in another part of the hotel?"

"I don't think so, Dad. I think she's just ancient, and no one wants to bother getting rid of her."

"She does consistently show up when she is scheduled, though, right?"

"Yes."

Caroline and I talk about why management would want to protect an underperforming team member. Caroline is convinced the reason Rhoda can get away with her restrictive schedule is because she has been there for a long time. I tell Caroline she could spend the rest of her high school years trying to find the answer about Rhoda, but it's best to let it go. Caroline wasn't so quick to let it go. The team talks about it all the time.

"If you did find the answer, then what?"

Caroline sighs, "I know, it doesn't really matter *why* it happens. But, like, it's not fair to everyone else. You and Mom have told us everything in life won't be fair, but this situation seems like it should be. Everyone else there knows that if they didn't, like, show up for holidays or weekends or training, they would be written up, and eventually fired."

"So, what does this situation do to the team?"

"It makes us hate her."

"Do team members target her?"

"No, because they know they'd get fired. But, like, everyone makes fun of her behind her back." Caroline sighs. "Dad, I try not to, but it's hard to resist. They make jokes all of the time."

"I can imagine," I picture a bunch of teenagers unhappy about being treated unfairly.

"Our team like, hates most of the managers for this, too, Dad. We, like, complain about them all of the time. Most of us will probably move on and get a different job. The managers probably know that, and, like, they don't even care. They know they will find more people like us, who are like, so happy to get our first job, we have no idea what is right or wrong."

"So, when managers hold onto an underperformer, the outcome is negativity on the team."

"It's a lot of things. But yeah, that's like, the big picture."

"It's interesting how this one employee can cause so much havoc on a team. All because she's given different boundaries. And it sounds like it doesn't bother her in the least."

Caroline rolls her eyes, "She acts like she owns the place. Honestly, Dad, she does absolutely *nothing* when she works. She just stands around, and then points out stuff for other people to do. I don't know how she does it. It has to be, like, so boring for her."

"Well, sounds like Rhoda is a permanent fixture at Splash. And, she is probably very tight with at least one of the managers or a general manager, so you should be very careful of that, Caroline."

"Yes. I should really stop joining in all of the jokes. It's just, like, so hard not to."

"Eventually, you may be bothered enough by this to look for a new job. I'm not saying you should or shouldn't; the grass isn't always greener somewhere else. But in this case, it might be. Work should be a place where you feel good about yourself. It's not ever going to be perfect, but you spend enough time there that it should provide mostly positive experiences."

"The only positive experiences I have are about making fun of all the junk that happens there. It's our culture."

"Imagine you are the manager of the water park, Caroline."

"You mean the water park director," she corrects me. One thing about kid number one…she is very precise.

"Right, imagine you have the job. You're new to it, and now you're telling the team about your expectations for working weekends, holidays, participating in the training, the customer is number one, all of that. What do you do when one of your employees tells you they have never had to work weekends or holidays with the former director? In other words, what if one, or more, of your employees resists your expectations? Keep in mind, you are under pressure from your leader to run a safe, fun, water park, in the smoothest way possible, with mostly teenage workers."

Caroline sighs, "Well, I wouldn't just let someone do it. It would be really hard, but I would like, have to meet with the person or persons, and like, talk about the expectations I had for them on the team. I guess I'd also have to like, talk to the general manager about the problem. I'd have to ask if I can fire them, if they don't change. You know, assuming they don't want to like, change."

"Okay, and what if you have those conversations, and the workers agree to your expectations, but then they don't follow through on them? They call in sick, come in late, whatever. What then, Caroline?"

"Ugh, I would have to fire them."

"Correct."

"But this doesn't happen at Splash. At least, I don't think it does. I think whenever there is a new manager, they just, like, inherit the problems of the last manager. It's easier to like, let people do whatever, than it is to make a change." Caroline sits back in her chair and sighs.

"It happens in large corporations too, Caroline. Many people are leaders who don't understand what they are getting into when they take on the responsibility to lead a team. Me too. At some moment, most of these leaders realize they need to do more with their job; more conversations with the leaders above them, more coaching and mentoring, and more courage to handle it all."

"But most probably don't care to like, do this," Caroline says. "At least, not the leaders of a hotel water park."

"Some do," I say. "I hope someday you find a job where this happens."

Caroline hugs me, and thanks me for my advice. She's not sure she will hang on at Splash much longer, but with school and basketball coming up, it's too much stress to go out and look for something new right now. I agree that focus is important. But I remind her that the leaders at Splash put off decision making indefinitely. Delaying the hard things doesn't make life easier. It may seem that way, until the next challenge further exploits it, and more problems develop.

"Caroline, your story is a really good example. One of the most difficult parts of being a good leader is knowing when to make a difficult decision, like letting go of an underperformer. And making sure you have the air cover of the leaders above you to confidently make the decision."

"That makes sense. Especially at Splash. I can't think of any good leaders there. I mean, I don't know all of them. But the ones I do know are not great. It's probably the same at the levels above them. It's probably a whole group of leaders who can't make good decisions and don't trust each other."

"They aren't the only ones."

Reflections

Good leadership takes time and energy.

Many leaders don't believe they have the time to lead.

Some senior leaders pressure mid-level leaders in ways that harm their teams.

Poor performers survive when leaders do not make boundaries for them...to grow or to move on.

When resentment builds, it becomes part of the culture.

Metrics Mania

The transformation team has just finished refining their backlog, when Thad shows up. He sits down in the back of the room, opens his laptop, and begins typing. It's as if he's there to observe, not participate. Karen notices him, and greets him. She has to do that, but I'm sorry she's doing it.

"I wasn't going to interrupt your meeting, but I have something important to share with you," the lizard face returns. "Is this a good time?"

Karen has no choice but to say yes. Well, she has a choice, but she doesn't perceive she has one.

"As I'm getting into my role as a scrum master, I've noticed some great inefficiencies demonstrated by the first scrum teams. There is a great opportunity for us to help them get off to a good start."

Karen gulps, "Can you please tell us what's on your mind?"

"I have attended several standups of these three teams." Thad is looking at the ceiling, searching. "I can't remember the names of the teams, but you know who I'm talking about."

Thad identified some great inefficiencies that he felt could be corrected with a little rigor around them. He uses his hands to make air quotes when he says the words *little rigor*. He has some ideas about how to clean up the waste on these teams, if we had time to listen. Great, Thad is going to lean out scrum. Does he know scrum's been around since 1996? I know the answer. And it's probably not worth telling him, either.

Thad lays out a plan to structure the dev team's day so that everyone doesn't have to be at every meeting they have. It's far too inefficient that way. The plan includes a calendar that's color coded, so it's easy to follow. He also made a checklist for the manager to follow, to make sure the team is doing what they are supposed to be doing.

The entire team is silent. And, uncomfortable. Thad is missing the point of the power of scrum: collaboration and the freedom to develop based on it.

"Thad, I have a few questions," Karen says with great trepidation. "I know you said you've observed our three scrum teams, but have you attended any scrum training, like Essence of Scrum?"

Good work, Karen. Find his baseline. Meet him in whatever twisted place he is at. Otherwise, agile at WL will be banjaxed by Rockin' Willy the Clown. Ugh. Now I have Thad as a clown in my head. Just so I don't mistakenly call him Rockin' Willy.

"I'm scheduled to attend Essence of Scrum training in a week. It's a special session for several of us from WL. But in the meantime, the waste is abundant and obvious to me. I don't need scrum training to see that."

"Can you please give us an example of the kind of waste you're talking about?" Karen is brilliant. I should take lessons from her on how to talk with him.

Thad sighs and searches his laptop, "I have a list, here, somewhere...Oh, here it is. Well, I mentioned the first one, about how all of the dev team members meet each day to talk about their work. Another one is how they talk to each other all day long. With them sitting in that open space so close together, it's a wonder they get anything done at all."

"Any other observations?"

Tongue sticking out, Thad squints at his list again, "There are many more, here. But let's move onto the sprint review, shall we?"

Karen nods and encourages him to share more. She's not faking her interest; she is sincerely trying to climb into Thad's head to find the meaning behind his ideas. His terrible, misguided ideas.

She's not only showing *me* a great example of leading change, she's showing our entire transformation team as well. They have already run into barriers and near-hostile situations; Karen's approach might be just the thing.

"Well, I observed the most recent sprint review. Their presentation wasn't very polished, and the actual demonstration was so slow. They shouldn't have been up there in front of everyone with such a painfully slow system. They really need to improve their flow for this event. Maybe it would have been better if a leader would have done it. You know, someone who really *knows the score*."

"Who knows the score, Thad?"

Thad chuckles, more to himself than anyone else. "That's a good question. A person with experience in public speaking. In this case, likely a leader who could make this presentation flow better. They would be able to get that system running faster. Or, they would have the perspicacity to hold off on showing the work until the system is running faster. People want to see the final product working. Not tottering along, like it's going to crash any moment."

Perspicacity? Rockin' Willy knows the score, and no one else. Man, what would Eve do in this case? Probably what Karen is doing. Seek to understand. But then she would offer coaching. Karen doesn't have the positional power to do that with Thad. I don't either, because I am not a coach. Well, Eve thinks I am, but no one at WL is ready for that. I keep thinking about what Eve would do...

Karen nods her head to show she's listening. In fact, it's some of the best active listening I've seen in a while. If the team didn't know her better, they might mistake her body language for agreement. Speaking of the team, they are mostly pale with surprise. Some of them are making furious notes on their tablets, to which Thad is oblivious.

"Did you provide the team with your feedback?" Karen asks.

"No. Actually, I'm perplexed on the overall value of the event. I understand customer focus very well, and it was good to see customers in attendance. But this work probably didn't need an event; it was very inefficient for all of us to stop working and attend. The team could have met with the customers on their own. A status report would have been good enough for everyone else. Or, maybe, this type of thing could be reported in the new C² Newsletter."

"C²?" Karen feigns curiosity.

"Oh, that's right. Not everyone received that email," Thad winces again. "There's a new culture committee forming at WL, to help us *grow happy, grow strong*. This type of event might be better off being distilled down into a report, and published in a newsletter."

"So, they should scrap the event all together?" Karen asks.

"Yes," Thad looks at his laptop. "I don't want to crush them so early in their journey, but it's so obvious this method isn't going to work at WL. We should stop it now, before these teams get too far down the road, and before this team spins up more teams. It will only lead to more problems."

"Do you have any other observations?" Karen asks. Her poise is steady. She's inspiring me to keep my emotions out of this. Without her fine example, I don't think I could sit with the team with a neutral face and my mouth shut. Also, she wisely avoids any more curiosity about C². No need for another distraction.

"Just one more, thank you," Thad looks to the ceiling, as if searching for the right words. He puffs his cheeks and blows out air through his mouth, like he's working up a really big comment. The whole room is rapt with attention, and Thad is soaking up every second of it.

"These teams are underperforming. I gathered metrics from each of the teams' scrum masters, and it's obvious they are really struggling. And, they are not consistent with each other. They are all over the place on their productivity."

Karen asks Thad about the metrics. He gathered the burndown charts from all three teams, for the last two sprints, which is four weeks of work. The teams are not getting done all that they committed to, and none of the team's burn down rates are the same.

Karen asked him if he discussed the data with the scrum masters. No, he hadn't, because he felt the situation was far reaching, and out of their hands. Thad felt it best to let the teams continue until he has a chance to bring these issues to senior management, so they can reconsider their investment. Thad also said there was a *seat at the table* of this discussion for us, because it directly impacts our work. Thad made this out to be a huge favor he was doing on our behalf.

Karen thanked Thad for sharing his feedback so candidly with us. The room still thick with silence, I am at a loss of what to say. Karen isn't.

"Thad, I can understand your concerns about the sprint review logistics, and the productivity of our new scrum teams."

Thad holds up a hand, and stops her, "I know what you're going to say," he grins at her for a half second before searching the ceiling for more inspiration. "I need to attend the training. And I will. But these concerns need to be addressed immediately. We can't let our good people be lead astray like this for long, or we will lose them. Then, everyone loses."

The laptop is closed, and Thad stands up. He thanks us, and leaves us, staring blankly at one another. Time for me to be a leader, even though I'm unsure of how exactly to do that.

"Well, that was an interesting experience, team."

Everyone sighs and talks quietly to each other. The team is really rattled by Thad's words. I can't blame them. I am, too.

"First and foremost, thank you, Karen for engaging Thad. You did all the right things to lead change: seek first to understand, actively listen, and thank them for sharing their feedback. You kept your emotions out of it, and stuck to the facts. Well, Thad's facts. It was an incredible thing to watch. Thank you."

The entire team claps and gives Karen all manner of team attention. She smiles, and tells us she is absolutely spent for the day. It was only about 15 minutes of conversation, but it was exhausting work. She said she has so many thoughts about the things Thad said, but mostly she is sad. Many team members said they were so glad it was her and not them. A disturbing, but true point.

"You gave us a great example of how to lead the conversation of change. It happens to be about agile and scrum, but what you did could be used for any kind of change. We are so fortunate to have witnessed the conversation."

"Thanks."

There are comments about Thad being clueless, him being a scrum hater, an agile hater, a people hater. They see him as egotistical, and no one missed the fact that Thad is unable to make eye contact with anyone for longer than one second.

The next question was, of course, what to do about Thad? Since he is going to senior leadership, the team felt this problem landed in my court. I agreed. And I need to do it fast. I encouraged the team to not spread rumors about Thad's comments. I realize this is irresistible to many, but I had to say it.

"The new scrum teams don't need more pressure than they already have," Karen adds.

"Yes," I sigh. "All right. Back to the change leader topic…"

I remind the team that all of us could be faced with a conversation like Karen had. Hopefully, that won't happen, but transformations impact people in different ways. Our team will face these conversations daily. Several people nod their heads, as they are experiencing a taste of this friction already. It's up to us to take the time with each conversation. If we don't, we will only have to backtrack later, and do more work. Or worse, we could lose people entirely. The conversations take time, and often will have to be repeated. The team suggested we will have to repeat the conversation several times with Thad.

"Let's hope not," Karen sighs. "I don't think I can handle much more."

The team disagrees, telling her she's fantastic, and that after a little rest, she will be ready to lead change again. I love seeing the team encourage one another.

"Joel, we all know most people who attend training have a much better understanding of scrum and agile. I'm not sure that's going to be the case with Thad, do you?"

"We could spend a lot of time speculating about it. And I understand, because he is a VP with a lot of influence at WL. The bottom line is, we really don't know, and may never know. Let's not spend any more team time on it. We have many other priorities to focus on."

"What if our work is all for nothing?" A team member asks.

I reassure the group that our senior leadership made an investment in agile. I remind them of the *grow happy, grow strong* vision, which includes changing how we deliver software to our customers. There are no guarantees, and the leaders could change their minds, but it's highly unlikely.

"What you are doing goes so far beyond leading a change to use agile and scrum. You are leading…learning. We are doing this because we are changing WL to be a learning organization. We do this by leading people in how to learn."

The team digests my words, and determines they need to discuss this further. Someone sets up a one hour discussion for tomorrow. They want to give themselves a night to sleep on it, and shake off the negativity of Thad's words.

"I know this might be hard to believe right now, but just because a VP gives you a bunch of negative feedback, it doesn't mean you are wrong. It's feedback. We need to listen to it, give it our attention, and even our empathy, but it's not a mandate or a decision. Feedback is feedback."

Listen to me. What do I know about any of this? And yet, the encouraging words are coming out of my mouth. Eve didn't coach me on this specifically, but our work together most certainly helped me be a leader in this moment.

"I get what you're saying about feedback, Joel. When the conversation began, I felt confident. I believed there was no way one person could drag down an entire transformation," Karen sighs. "I'm no longer confident of that. This guy seems capable of it all."

Reflections

Leading change takes a lot of energy and time, just like good leadership.

Meeting people where they are neutralizes the conversation.

Active listening helps meet people where they are, and shows authenticity.

Using the former process metrics of waterfall can make agile and scrum look unproductive.

Thad, and people like him, will steal all of our energy if we allow it.

My team needs leadership, support and encouragement.

Training Frenzy

It's late in the afternoon when I return to my desk. I have to pack up soon because Elliot has a basketball game. Three kids in basketball is a lot of games. Cele and I can't attend them all, so we often try to divide and conquer. This is great, but then Cele and I see even less of each other. So we've created the Game of the Week, where both of us attend the game together. We can talk, watch the game, and just…relax.

Eve's coaching is one of the reasons I even have a strategy to see any of my kids' events. Making time for what matters most to me is part of the original vision I made for myself, personally and professionally. It's still not perfect but I have actual balance, now. And the confidence to uphold the balance when things get rocky.

Marilyn walks into my office, "You've been gone so much."

"Yes. My new way of working takes me to people, instead of people coming to me. It's rather refreshing. But I imagine it drives you nuts."

Marilyn rolls her eyes in play, "You know it. Listen, you and your team must be doing something different this week."

"Why?"

Marilyn tells me I have urgent calls from HR, corporate planning, information risk management, and of course, facilities. That doesn't begin to touch all of the emails waiting for me. She also scolded me for missing a strategic expense management meeting. Ugh.

"Huh. What do all of these callers want?"

"I've never seen anything like it, Joel. They are acting like you have the answer to everything. You went from being under the radar to being the answer to everything."

"Yikes."

"So, you're not the answer to everything? You haven't been secretly promoted or anything like that?"

"Nope. People are learning about the transformation team. I reckon they are reacting the way we always do with change at WL. There is a project outcome manager or leader, and everyone skips over the teams doing the actual work to ask that person what's going on. I think that's what is happening here."

"Okay, so no promo?"

"No, Marilyn. That has to be the reason. Our team has begun training others, and hosting events to share the vision and talk about the changes. I'm not doing most of that work, but people are used to going to the leader. They don't know it yet, but my entire transformation team is full of leaders."

"But…they all want to know where they can get training. Some of them want to know how they can help, and some of them are mad at you because one of your team members gave them feedback. Oh, and the weirdest call of the day: someone named Chris, who wants to make a follow up appointment with you on why agile fails."

"Whoa. They must not see our new transformation website. And the mad person, yeah, I need to talk with them. And yes, Chris the scrum master. I do need to meet with him again. In the next week, if he can."

"They didn't. I direct them to it, but they just want to talk to you. Isn't that your responsibility, to talk to them?"

"It is, but the entire transformation team also has this responsibility. They know they can handle any question from anyone, and that I'm here to help. And, thank you for directing people to the website. We have a ton of basic information there…and a training schedule."

"But…what exactly are you doing, Joel? I don't get it."

"I have a responsibility to sponsor the team and work with our stakeholders."

I pause.

"Marilyn, what do you think about starting to attend our daily standup meetings?"

"To take notes and look for trends?" She asks.

"No, I want to do a better job of bringing you along on this change. It's for everyone. Especially someone like you, out here on the front lines every day," I sigh, mad at myself for not thinking of this until now. "Marilyn, would you like to join the transformation team? Please say yes."

She grins, "Really? I would love it. Yes!"

"We should have done this earlier, but I'm really learning this job as I go, Marilyn."

"That's okay. But...how will I still do my job to support you and Vijay and the team?"

"They are on the team. It shouldn't be extra work, just...different."

"What will I need to do?"

"The team will help you, but something tells me your initiative will be a natural fit for our team. You will sense, adapt, and adjust, just like I do. Learn how to continually learn, and show others how to do it, too."

"I'm willing to give it a try."

"Great. I can use all the help I can get," I sigh.

Marilyn sighs, and shakes her head at me, "You, my friend, are *Hunda P* in the eye of the tiger right now. You need my help, I'm here for ya."

"Hunda P?"

Marilyn casually shrugs, "It's the hip way to say 100 percent, Joel. You are 100 percent in the eye of this tiger."

I laugh, "You're not kidding."

Reflections

One of my biggest supporters needs can be an active member of the transformation team, instead of a supporter on the sidelines.

The change I'm leading is beginning to happen.

I am hunda-P in over my head.

My Vision: Reloaded

"Joel, today we are going to revisit your vision," Eve pulls out a document of my vision, like she does at the start of all of our meetings. "We've had so many hot topics lately, we haven't reviewed your vision in quite a while.

My Vision

Professional
- To be trusted by Lora and Rick.
- Less wasted time on conference calls.
- Fewer ultimatum days: You can do this, but then you won't have that.

Personal
- Eat better.
- Support my kids' events. I miss too many of them.
- Fewer ultimatum days: You can do this, but then you won't have that.

"We're only doing this as part of the standard process, right? You're not dumping me, are you?"

Eve smiles, "Going to the negative right away, Joel?"

"Yeah, I guess I am. I've just been in this…this *mode* lately. But I did reflect on this vision."

"Well, let's look at your vision. Who knows, it may help you get out of your mode. And, we'll see where *how do you lead learning?* fits into it. That was the other part of the assignment."

We discuss my most important priority, to be trusted by Lora and Rick. With Lora out of the equation, how does this priority feel? I am one hundred miles from that right now, and I definitely don't need it as my number one priority anymore. Rick and I are definitely not in a perfect relationship, but we are also not in the highly dysfunctional place we were when Lora was still here. I've made progress on the goal thanks to Lora leaving, and thanks to Eve's coaching. The value and calendar work I did with Rick and his peers is a good measure of progress.

So much of that situation was Lora and the environment she created around her; I wasn't doing untrustworthy things. In fact, it was the opposite. And the worst part of it is how I was made to feel that I couldn't be trusted.

Lora led by intimidation, and had her hands all over our work. Eve called her a classic command and control leader; I called her miserable. Often, she would change things without my knowledge, or other directors reporting to her. This is how Meryll and I became such good friends. We had to meet almost daily to strategize on how to be successful, despite Lora. Our meetings were full of speculation on what Lora might do; if she does this, then we do that…Then when Jack was hired, he quickly folded into our group. We became the three amigos, united in our quest to survive Lora's reign.

The next priority, less wasted time on conference calls, is also obsolete. My days are no longer sitting through a string of 1x1s in my office and endless conference calls. I work with Marilyn on my calendar to find the valuable interactions within my week. She and I have a great cadence in doing this and it helps us work better together.

I changed and grew by practicing the lean leader concept *Go & See*. These days, I'm not in my office very much because I'm visiting my teams, and talking with people who need to hear about WL's transformation. I *do* the work with the teams and stakeholders; I don't just talk about it and take it back to my desk to do it alone. We discuss, solve problems, and work in real time. It's refreshing and energizing.

I also have a healthy practice of making sure my calendar has balance. I'm no longer working late, on weekends, so I'm able to see more of my family.

It's not perfect, but I have clarity each day, and less mental conflict about what I should be doing.

"What about *fewer ultimatum days?*" Eve asks.

I sigh, "That one is still an issue, but in a different way."

Eve wanted to know more details. I tell her that things feel very chaotic right now. With another team that seems to be competing with us, I find myself second guessing what I should be doing. If I do this, what will Thad do? Should I do this so that Thad doesn't do that…

"Is it any better since we last discussed it?"

"Yes. And when I look at the statement ultimatum days, it's very related to this. But I think it's more ultimatum moments. I shouldn't be strategizing around Thad's manipulation every moment of the day."

"Okay. Joel, this goal was originally about how you spent your time. You felt backed into a corner because there were so many meetings, and you were conflicted about where to spend your time."

"That's true. I've grown out of that. The calendar clean up, practicing *Go & See*, and just focusing on value has helped me grow out of that problem."

"Excellent," Eve says.

"This other stuff with that sounds different now. Second guessing."

Eve nods, "We'll get to the new goals next. I have a note here, so we don't lose it. Now, onto your personal goals. That first priority was to eat better. We didn't really work on this, but we can talk about it if you want. Although, when I saw your energy of choice while out on the Ironman Boulder course, I'd say you are not hitting this goal."

I chuckle, "Good point. Well, that day, all the usual ways of surviving were off the table. But even on regular days, I still want to work on that one. I was doing well for a while, but then this Thad problem has derailed me. I find myself eating faster, and thinking about Thad while I'm eating. I have to stop that."

"Fair enough," Eve says. "We'll talk about that one in a moment, too."

"Onto that next one, *supporting my kids' events*, I'm really proud of what I've accomplished there. I see at least one of their games each week. It feels really good, Eve."

"And you have a Game of the Week, right? Where you and Cele attend together?"

"Yes. The changes I've made professionally have made a very positive impact on my entire family. And, I no longer have so much noise in my head about choosing between WL and my kids' events."

"Let's take that as a win," Eve makes a note.

We discuss my last goal, which is the same for me professionally as it is personally. Although I felt better about the changes I've made through coaching, I didn't realize how much I had changed until this moment. I used to feel guilty about missing the kids' events and being at work, and guilty about being at my kids' events and not being at work. I was fearful of Lora coming down (even harder) on me than I already endured.

The other measure of success for this goal was that I was able to train for a successful finish at Ironman Boulder. It's all because of Eve's coaching that I was able to first believe I had the time to train for and Ironman, and then to actually do it. It wasn't easy, and took strategy, but I did it. What a fantastic feeling.

"What feels the best, Joel?"

"What do you mean?" I think I know what she means, but it's a huge question.

"Of all the growth and change we discussed, what feels the best?"

I sigh, "That's tough. It's two things: supporting my kids' events, and less wasted time. They are very connected."

"Joel, what bothers you the most right now? It can be one of these goals, or something else."

"It's got to be about leading learning. I refuse to make dealing with Thad my number one goal."

"Why? It seems like the way he works is really affecting you."

"It is, Eve. But he is one person. We have an entire organization to transform."

"I like that you said *we* and not *I* regarding transforming the organization."

"Thanks. We have a great team on it."

"So what is the number one goal right now? Transforming an organization is too big."

"Yes," I pause.

"Not forever. Just for right now."

"I want to lead learning with my team," I say.

Eve looks satisfied, but waits for me to say more.

"If I can lead learning with my team, they can carry it forward in their interactions. It might not be exactly the same as the way I do it, but they need to bring it to the organization."

"This goal sounds big, too, Joel. Do you think it's the right size?"

I pause, "Uh, it is pretty big. And it's more like how I'll work, than an actual goal."

Eve and I try to break the goal down further. I decide that leading learning really is too big. The goal I want for my team is to lead them in *practicing learning*. This means encouraging experiments, giving air cover when the experiment goes bad or when the team makes a mistake; encouraging reflection, encouraging feedback, and helping them get comfortable adapting to a new direction. Wow.

"That's a good goal, Joel. Help them practice learning," Eve says. "Then, they can go out and help others practice it, too."

Eve writes this goal in the first priority slot, and tells me we will dig into this in our next session. Of course we will. And she'll probably make me do something uncomfortable to practice it, like help Lilly and Jessica with their restaurant or café.

"Next, I need to end the methodology war."

Eve nods, "There will always be tension in a transformation, but there should not be wars. No one wins in those."

"Yeah, there is a lot of chaos right now. I can't wait to dig into this goal."

"As you dig in, I will encourage you to research why agile isn't just a methodology."

"It's a mindset."

"Yes. The war you're fighting conveniently skips right over that fact."

"Roger that." I say it as if I know what I'm doing with that goal. I don't. I need Eve's help. And, my team's help.

"What else? We've got *practice learning with transformation team* and *end methodology war*. Is there another big professional goal, here?"

I sigh, "Uh…that's already a ton. Can I just have two professional goals? If I can accomplish these two in the near future, I would be ecstatic."

Eve smiles, "These are big. Let's hold it at two, and see how it works. What about personal goals?"

I know exactly what my number one personal goal is: get Thad out of my head. Only after doing that can I eat better and have a quiet mind again.

And there, I'm at three very connected personal goals: get Thad out of my head, eat better and regain mental clarity.

Eve shows me my new vision:

My Vision

Professional
- Practice learning with transformation team.
- End methodology war.

Personal
- Get Thad out of my head.
- Eat better.
- Regain my mental clarity.

We take a moment to read what we've built. In just a few moments, it's so clear on paper. So easy to see now. There has been so much happening around me and the team that it felt so complicated and out of control. That's coaching for you. Making sense in the chaos.

"How do you feel about it, Joel?" Eve asks.

I raise my eyebrows, "Impressed with what you got me to commit to."

Eve waits.

"It feels unlikely that I can end the methodology war, rather than tone it down to a small battle that only flares up once in a while. But I know that it has to be completely ended, or we'll face fractioned adoption and successes."

"Unlikely," Eve smiles. "Ending the war is most definitely unlikely work. There will be a series of very hard moments and decisions that will stack up over time to either success or something less desirable than that. It's nuanced work, so we'll have to watch this goal closely."

Eve asks me for more thoughts.

"I'm leading the unlikely again."

"Yes, you have been since the very start. New challenges. New unlikely challenges to lead."

I lean back and sigh. Eve looks at me, waiting for more.

"I'm definitely uncomfortable with the difficulty, here. And the effort needed. But it's very exciting to think that I can make this kind of change. I believe I can do it. Well, with my team."

I tell Eve the excitement I feel goes beyond the thought of accomplishing the goals: it's about a better quality of life, which will make me a better leader, a better father, a better husband. Even though I've made great progress to be present with my family, I've been slipping back to WL mentally. I hate that I've allowed Thad to do that to me. Enough is enough; now I have a plan to change it. For good.

Ending the methodology war will help me stop overthinking anything to do with Thad, which will get the noise out of my head, and I'll stop stress eating crappy food. This will help with my Ironman Boulder training. Just not the one The Mouth does. Eve asks if I should put Ironman as a goal. I didn't think it needed to be a goal like these. Ironman is icing on the cake, not something I must do to become a better leader or person.

"So, you remember this from the last time we built a vision. You have a 24 hour free-look period, where if you have any qualms or questions about it, call me and we'll talk. Otherwise, we're off and running on coaching to these goals."

"Yes, I remember that."

"The next time we meet, we'll talk about what's getting in the way, and then we'll define a breakthrough project for you to work on. This is how you will chip away at your goals, one small chunk at a time. I'm excited to keep working with you, Joel. There will come a day when you will get to a level leadership and agile mindset that you do not need me. For now, let's tackle these unlikely, uncomfortable, energizing goals together. Let's tackle growth."

I smile. Leading unlikely, I am uncomfortable and energized. I wouldn't have it any other way.

To be continued...

Join Joel, as he continues his journey to become an agile leader. While WL wades through a competitor disruption and the fallout of a major security breach, talk of layoffs are more frequent. Rumors say that the corporate employee engagement initiative *grow happy, grow strong* is actually a Trojan horse for mass layoffs, and lean will be used to make it happen. What is the truth, and what is just a rumor?

The transformation team and Joel struggle to find the truth as Joel leads them to end the methodology war started by Thad. Joel's refreshed goals have him energized to take on the work; will it be enough to end the war, and focus on transformational change?

Author Biography

A conqueror of mediocrity, Francie Van Wirkus is a certified professional coach for executives, product managers, and teams. She works in the insurance and financial services industry, with an Agile, Lean and sales focus. She helps people of all levels take strategic, sustainable steps toward their goals. Herr innovative ways of leading and sustaining change help lead IT transformations and personal transformations of leaders; she is a SAFe 4.0 Program Consultant.

Francie is a change thought leader, speaking to groups and coaching them to continuously learn and adjust. She is co-founder of the Agile Bettys, a podcast and consulting firm that helps people live an agile lifestyle.

She is author of the travelogue memoir series The Competitor in Me, and a business novel series on leadership, The Dominant Gene.

Francie enjoys teaching Bikram yoga, racing in Ironman competitions, and lives in Wisconsin with her husband and three delightful children.

www.francievanwirkus.com
www.agilebettys.com
www.ab2consultingservices.com

Email
francievanwirkusinspire@gmail.com
ab2consultingcoaches@gmail.com

Francie can also meet you on Linkedin, Twitter, and Instagram.

Enjoy Other Books by Francie

The Competitor in Me memoir series, exclusively on Amazon Kindle

Through memoirs, The Competitor in Me is in inspirational travelogue, engaging the reader in a poignant story of how each of us, with perseverance, can do what seems impossible. Breaking through requires conquering yourself many times over, by winning epic battles with doubt demons, who constantly work to make you feel less than the amazing person you are.

Catholic Mother Problems, exclusively on Amazon Kindle

A comedy of family errors, a tale of two mother's guilt. When two mothers steeped in Catholic tradition discover their children's marriage is in jeopardy, they decide they're not going to take it lying down…